STUDENT SOLUTIONS MANUAL

to accompany

Discrete Mathematics

Mathematical Reasoning and Proof with Puzzles, Patterns, and Games

Douglas E. Ensley
Shippensburg University

J. Winston Crawley
Shippensburg University

WILEY

JOHN WILEY & SONS, INC.

Cover Photo: ©Chuck Carlton/Index Stock.

To order books or for customer service, please call 1-800-CALL-WILEY (225-5945).

ISBN-13   978-0-471-77367-2
ISBN-10   0-471-77367-0

Printed in the United States of America.

SKY10078574_062724

Printed and bound by Quad/Graphics.

CONTENTS

Chapter 6

Chapter 7

Student Solutions Manual

Discrete Mathematics: Mathematical Reasoning and Proof with Puzzles, Patterns, and Games

by Douglas E. Ensley and J. Winston Crawley

Section 1.1 exercises

1. For each of the games, we show all ten steps in the trick:

(a) $C(1,1,1,\text{no})$	(c) $C(3,3,3,\text{no})$
1. HCDs	1. HCDs
2. sHCD	2. CDsH
3. sHdc	3. CDhS
4. csHd	4. DhSC
5. csDh	5. Dhcs
6. hcsD	6. hcsD
7. hcsD	7. hcsD
8. hcsd	8. hcsd
9. hcDS	9. hcDS
10. hsdC	10. hsdC

3. In the Josephus game ...

    (a) ... $J(15,3)$, person 5 is the last one and person 14 is the next-to last-one.

    (c) ... $J(15,2)$, person 15 is the last one and person 7 is the next-to last-one.

4. The game tree is shown below. Note that 12 of the 20 total outcomes are five-set matches. Does this seem like a reasonable proportion?

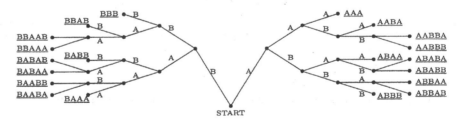

7. (a & b) The game tree is shown below, with the first branch representing the nickel, the second branch the dime, and the third branch the quarter. For example, HTT means the nickel is "heads" and the dime and quarter are "tails." The "systematic" list appears on the right.

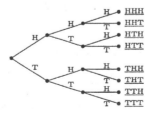

    (c) There are three outcomes where two "heads" were tossed and only one outcome where three "heads" were tossed, so it seems that getting exactly two "heads" is more likely than getting three "heads."

9. Relabeling "H"s in Problem 7 with "F"s makes the game tree identical to the one in Problem 8.

11. Refering to Figure 1.2, all ways of drawing the envelope must start at corner 1 and end at 2, or start at 2 and end at 1.

12. For games (a) and (c), there is a move (highlighted below) that will guarantee a win. In (b), you will lose if your opponent knows what she is doing.

(a)

X	☒	X	X
X	X	X	
X		X	X
X	X	X	X

(c)

☒		X	X
X	X	X	
X		X	X
X	X	X	

Section 1.2 exercises

1. (a) Next term is 14, and $a_n = a_{n-1}+2$ with $a_1 = 2$, or $a_n = 2n$; (c) Next term is 50, and $a_n = a_{n-1}+(2n-1)$ with $a_1 = 2$, or $a_n = n^2 + 1$; (e) Next term is 64, and $a_n = 2 \cdot a_{n-1}$ with $a_1 = 1$, or $a_n = 2^{n-1}$; (g) Next term is 20, and $a_n = a_{n-1} + 3$ with $a_1 = 2$, or $a_n = 3n - 1$; (i) Next term is 320, and $a_n = 2 \cdot a_{n-1}$ with $a_1 = 5$, or $a_n = 5 \cdot 2^{n-1}$; (k) Next term is 25,000, and $a_n = a_{n-1} \cdot a_{n-2}$ with $a_1 = 2$ and $a_2 = 5$; (m) Next term is 18, and $a_n = a_{n-1} + 3$ with $a_1 = 3$, or $a_n = 3n$;

2. (a) $a_{k-1} = 5(k-1) - 2 = 5k - 7$, and $a_{k+1} = 5(k+1) - 2 = 5k + 3$; (c) $a_{k-1} = 2(k-1) + 7 = 2k + 5$, and $a_{k+1} = 2(k+1) + 7 = 2k + 9$; (e) $a_{k-1} = 2^{3(k-1)+1} - 1 = 2^{3k-2} - 1$, and $a_{k+1} = 2^{3(k+1)+1} - 1 = 2^{3k+4} - 1$;

4. (a) $a_n = 2 + a_{n-1}$ with $a_1 = 2$; (c) $a_n = 1 + 2a_{n-1}$ with $a_1 = 2$.

5. (a) $2, 4, 6, 8, 10$; $a_n = 2n$; (c) $2, 5, 11, 23, 47$; $a_n = 3 \cdot 2^{n-1} - 1$;

6. (a) If $a_n = 2^n - 1$, then $a_{n-1} = 2^{n-1} - 1$, so $2 \cdot a_{n-1} + 1 = 2(2^{n-1} - 1) + 1 = 2^n - 1$. (c) If $a_n = 5n + 3$, then $a_{n-1} = 5n - 2$, so $5 \cdot a_{n-1} - 3 = 25n - 13 \neq 5n + 3$.

7. (a) $4, 7, 10, 13, 16$; $a_n = a_{n-1} + 3$ with $a_1 = 4$; (c) $9, 11, 13, 15, 17$; $a_n = a_{n-1} + 2$ with $a_1 = 9$; (e) $1, 8, 15, 22, 29$; $a_n = a_{n-1} + 7$ with $a_1 = 1$;

8. (a) $5, 25, 125, 625, 3125$; $a_n = 5^n$; (c) $3, 6, 12, 24, 48$; $a_n = 3 \cdot 2^{n-1}$; (e) $1, 5, 9, 13, 17$; $a_n = 4n - 3$;

9.   (a) The first 5 terms are $5, 5 + 6 = 11, 11 + 7 = 18, 18 + 8 = 26, 26 + 9 = 35$. If we don't simplify, they are $5, 5 + 6, 5 + 6 + 7, 5 + 6 + 7 + 8, 5 + 6 + 7 + 8 + 9$. We write the n terms that make up a_n in increasing order and in decreasing order as in the example:
$$a_n = 5 + 6 + \cdots + (n+3) + (n+4)$$
$$a_n = (n+4) + (n+3) + \cdots + 6 + 5$$
Adding, then dividing by 2 (as in the example) yields the closed formula $a_n = \frac{n(n+9)}{2}$.

11. $a_n = 99 + 2n$

13. $a_n = (2n - 1)^2$

15. When $a = 2$ and $b = -1$, the equation $(a+b)^3 = a^3 + 3a^2b + 3ab^2 + b^3$ becomes $(2-1)^3 = 2^3 + 3(2)^2(-1) + 3(2)(-1)^2 + (-1)^3$, or more simply
$$1 = 2^3 - 3 \cdot 2^2 + 3 \cdot 2 - 1$$

17. The nth term is $a + 3(n - 1)$

19. We might conjecture that $s_n = 2 \cdot a_n - 2$ from the table below:

n	1	2	3	4	5	6
a_n	2	4	8	16	32	64
s_n	2	6	14	30	62	126

21.   (a) $3 + 6 + 9 + 12 + 15 + 18 + 21 = 84$
      (c) $4 + 4 + 4 + 4 + 4 + 4 + 4 + 4 + 4 = 36$

23. (a) $s_5 = \sum_{k=1}^{5} 3 = 3+3+3+3+3 = 15$, and $s_{10} = \sum_{k=1}^{10} 3 = 3+3+3+3+3+3+3+3+3+3 = 30$

24. (a) $\sum_{k=1}^{9} 4k$

(c) $\sum_{k=1}^{6} 5$

27. (a) i. Anne (she is in position 1)

ii. The person with name tag 5 (he is in position 1).

(b) i. Player 4

ii. Players 5, 6, 1, 2, and 3 are left, in that order.

iii. Player 5 will win.

Section 1.3 exercises

1. (a) Only B is telling the truth, as we see from the following table:

	p	q	A says $p \wedge q$	B says $\neg p$
	T	T	T	F
	T	F	F	F
$\star$	F	T	F	T
	F	F	F	T

(b) Only A is telling the truth for sure, and we also know that exactly one of B or C is truthful, but we cannot tell which.

	p	q	r	A says $\neg q \vee \neg r$	B says $\neg r$	C says $p \wedge r$
	T	T	T	F	F	T
$\star$	T	T	F	T	T	F
$\star$	T	F	T	T	F	T
	T	F	F	T	T	F
	F	T	T	F	F	F
	F	T	F	T	T	F
	F	F	T	T	F	F
	F	F	F	T	T	F

3. (a) $\neg p \vee q$; (b) $p \wedge q$; (c) $p \wedge \neg q$

5. (a) $f \wedge (\neg m)$

(c) $(\neg f) \wedge (\neg m)$, or $\neg(f \vee m)$

6. (a) This person is a male math major.

(c) This person is a female, and she is either over age 30 or a math major.

7. (a) $t \wedge d \wedge h$

(c) $(t \vee h) \wedge \neg(t \wedge h)$, or $(t \wedge (\neg h)) \vee ((\neg t) \wedge h)$

8. (a) Bill is tall or dark or handsome, but not all three.

(c) Bill is dark, but not tall and light.

9. (a) $p \wedge \neg s$

(c) $\neg p \wedge (s \vee r)$

11. (a)

p	q	$\neg p$	$\neg p \vee q$	$p \wedge (\neg p \vee q)$
T	T	F	T	T
T	F	F	F	F
F	T	T	T	F
F	F	T	T	F

p	q	r	$q \vee r$	$p \wedge (q \vee r)$
T	T	T	T	T
T	T	F	T	T
T	F	T	T	T
T	F	F	F	F
F	T	T	T	F
F	T	F	T	F
F	F	T	T	F
F	F	F	F	F

(d) is indicated to the left of this table.

12. (a) Let h and t represent the same statements as in the example, and p represent "the snackbar makes a profit". Then the statement can be written, $\neg h \vee (t \wedge p)$, and it has the following truth table:

h	t	p	$t \wedge p$	$\neg h \vee (t \wedge p)$
T	T	T	T	T
T	T	F	F	F
T	F	T	F	F
T	F	F	F	F
F	T	T	T	T
F	T	F	F	T
F	F	T	F	T
F	F	F	F	T

(c) Let f and p represent the same statements as in the previous problem. Then the statement can be written, $\neg p \wedge f$, and it has the following truth table:

f	p	$\neg p$	$\neg p \wedge f$
T	T	F	F
T	F	T	T
F	T	F	F
F	F	T	F

13. In each case we apply DeMorgan's laws and the Double Negative law to simplify the negation. For simplicity, we apply the Double Negative law without showing the intermediate step.

(a) $\neg(\neg h \vee (t \wedge p)) = h \wedge \neg(t \wedge p) = h \wedge (\neg t \vee \neg p)$. Everyone is hungry at mealtime, and either everyone is not tired or the snack bar does not make a profit.

(c) $\neg(\neg p \wedge f) = p \vee \neg f$. The staff is well paid, or they are not friendly.

14. $(b \geq 600) \vee (m \geq 25)$. The negation is $(b < 600) \wedge (m < 25)$.

16. The equivalence follows from the fact that the columns of the truth table in bold italics are identical.

p	q	$p \wedge q$	$\neg(p \wedge q)$	$\neg p$	$\neg q$	$\neg p \vee \neg q$
T	T	T	*F*	F	F	*F*
T	F	F	*T*	F	T	*T*
F	T	F	*T*	T	F	*T*
F	F	F	*T*	T	T	*T*

18. The equivalence follows from the fact that the columns of the truth table in bold italics are identical.

p	q	r	$q \vee r$	$p \wedge (q \vee r)$	$p \wedge q$	$p \wedge r$	$(p \wedge q) \vee (p \wedge r)$
T	T	T	T	*T*	T	T	*T*
T	T	F	T	*T*	T	F	*T*
T	F	T	T	*T*	F	T	*T*
T	F	F	F	*F*	F	F	*F*
F	T	T	T	*F*	F	F	*F*
F	T	F	T	*F*	F	F	*F*
F	F	T	T	*F*	F	F	*F*
F	F	F	F	*F*	F	F	*F*

21. In each part we use bold italics to highlight the columns that must be checked to see if the statements are equivalent.

(a)

p	q	$\neg p$	$\neg p \vee q$	$p \wedge (\neg p \vee q)$	$\neg p$	$p \vee q$	$\neg p \wedge (p \vee q)$	
T	T	F	T	***T***	F	T	***F***	
T	F	F	F	***F***	F	T	***T***	(Not equivalent)
F	T	T	T	***F***	T	T	***T***	
F	F	T	T	***F***	T	F	***F***	

(c)

p	q	$\neg q$	$\neg q \wedge p$	$\neg p$	$\neg p \wedge q$	$(\neg q \wedge p) \vee (\neg p \wedge q)$	$\neg p$	$\neg q$	$\neg p \vee \neg q$	
T	T	F	F	F	F	***F***	F	F	***F***	
T	F	T	T	F	F	***T***	F	T	***T***	(Not equivalent)
F	T	F	F	T	T	***T***	T	F	***T***	
F	F	T	F	T	F	***F***	T	T	***T***	

23. (a) $\neg(\neg p \vee \neg q)$ is equivalent to $\neg(\neg p) \wedge \neg(\neg q)$ by DeMorgan's Law, and this is equivalent to $p \wedge q$ by two applications of the double negative property.

(b) $\neg(\neg(\neg p))$ is equivalent to $\neg p$ by the double negative law

24. (a) $(p \wedge \neg q) \vee p \equiv p \vee (p \wedge \neg q)$ by Commutative property, and $p \vee (p \wedge \neg q) = p$ by Absorption property

Section 1.4 exercises

1. (a) $(x > 0) \wedge (y > 0)$

(c) $((x > 0) \vee (y > 0)) \wedge \neg((x > 0) \wedge (y > 0))$

3. (a) $2, 4, 6, 8$

(b) $6, 7, 8, 9$

(c) $6, 8$

(d) 4

(e) $3, 6, 9$

(f) $2, 5$

5. (a) $2, 4, 6, 8, 10$

(b) None of the elements of D. That is, the predicate $R(n)$ is true for all the elements of D.

(c) $2, 4, 6, 8, 10, 12$. (that is, all the elements of D)

(d) 2

7. (a) is false since $x = 1 \in D$ is not even; (b) is true; (c) is false since $x = 1 \in D$ is odd; (d) is false since $x = 8 \in D$ is not odd.

8. (a) $\forall s \in B, G(s)$, where B is the set of biology majors and $G(s)$ is the predicate, "s is required to take geometry."

(b) $\exists s \in C, \neg M(s)$, where C is the set of computer science majors and $M(s)$ is the predicate, "s minors in mathematics."

(c) $\forall s \in M, \neg B(s)$, where M is the set of math majors and $B(s)$ is the predicate, "s is required to take a business course."

(d) $\exists x \in P, \neg S(x)$, where P is the set of puzzles and $S(x)$ is the predicate, "x has a solution."

10. (a) and (d) have the same meaning

(a) $\neg \exists s \in B, G(s)$

(b) $\exists s \in B, \neg G(s)$

5

(c) $\forall s \in B, G(s)$

(d) $\forall s \in B, \neg G(s)$

11. (a) Let D be the set of all friends of Alaina, and $P(x)$ the predicate, "x gets tired of playing at the beach." The two forms are "$\neg \exists x \in D, P(x)$" and "$\forall x \in D, \neg P(x)$".

(b) Let D be the set of all friends of Alaina, and $C(x)$ the predicate, "x dislikes doing cartwheels." The two forms are "$\neg \exists x \in D, C(x)$" and "$\forall x \in D, \neg C(x)$". (or we could let $L(x)$ stand for "x likes doing cartwheels" and write "$\neg \exists x \in D, \neg L(x)$" and "$\forall x \in D, L(x)$").

(c) Let M be the set of all math courses, and $T(x)$ the predicate, "x is too hard for Jennica." The two forms are "$\neg \exists x \in M, T(x)$" and "$\forall x \in M, \neg T(x)$".

(d) Let M be the set of all meals at the camp, and $B(x)$ the predicate, "x is too bad." The two forms are "$\neg \exists x \in M, B(x)$" and "$\forall x \in M, \neg B(x)$".

13. (a) (i) -12 (ii) $0, 2$ (iii) any even number is a counterexample

(b) (i) $-3, 23, 3, -31$ (ii) use $3 - 2y$ where y is the number they have chosen

14. (a) $\neg(\forall a \in \mathbb{R}, \forall b \in \mathbb{Z}, a^2 + b \in \mathbb{Z})$ converts to $\exists a \in \mathbb{R}, \neg(\forall b \in \mathbb{Z}, a^2 + b \in \mathbb{Z})$, then to $\exists a \in \mathbb{R}, \exists b \in \mathbb{Z}, \neg(a^2 + b \in \mathbb{Z})$, finally to $\exists a \in \mathbb{R}, \exists b \in \mathbb{Z}, a^2 + b \notin \mathbb{Z}$

(c) $\neg(\forall x \in \mathbb{Z}, \exists y \in \mathbb{R}, x = 2y)$ converts to $\exists x \in \mathbb{Z}, \neg(\exists y \in \mathbb{R}, x = 2y)$, then to $\exists x \in \mathbb{Z}, \forall y \in \mathbb{R}, \neg(x = 2y)$, finally to $\exists x \in \mathbb{Z}, \forall y \in \mathbb{R}, x \neq 2y$

17. (a) There is an integer x that is at least as big as every integer.

(b) There is a set of integers that does not have a smallest number.

(c) There is a positive integer x such that no matter how the positive integer y is chosen, either y is at least as big as x or y is not a factor of x.

18. (a) original statement (choose $y = x + 1$) (b) negation (the set of all negative integers has no smallest number) (c) negation (choose $x = 1$)

Section 1.5 exercises

1. (a) $\neg c \to f$

(c) $l \wedge \neg b$

(e) $\neg(t \wedge s)$

2. The truth tables follow below.

(a) False only if you don't attend the concert and you don't get an F.

(c) True only if I ate lunch but not breakfast.

(e) False only if this triangle has both a thirty and a sixty degree angle.

	c	f	$\neg c \to f$
(a)	T	T	T
	T	F	T
	F	T	T
	F	F	F

	l	b	$l \wedge \neg b$
(c)	T	T	F
	T	F	T
	F	T	F
	F	F	F

	t	s	$\neg(t \wedge s)$
(e)	T	T	F
	T	F	T
	F	T	T
	F	F	T

	p	q	$p \wedge q$	$(p \wedge q) \to q$
4. (a)	T	T	T	T
	T	F	F	T
	F	T	F	T
	F	F	F	T

6

p	q	$p \vee q$	$(p \vee q) \to q$
T	T	T	T
T	F	T	F
F	T	T	T
F	F	F	T

p	q	r	$q \to r$	$p \wedge (q \to r)$
T	T	T	T	T
T	T	F	F	F
T	F	T	T	T
T	F	F	T	T
F	T	T	T	F
F	T	F	F	F
F	F	T	T	F
F	F	F	T	F

(e) corresponds to the second table above.

5. In each part we highlight the columns that must be checked to see if the statements are equivalent.

(a)

p	q	$p \to q$	$q \to p$
T	T	T	T
T	F	F	T
F	T	T	F
F	F	T	T

(Not equivalent)

(c)

p	q	$p \to q$	$p \wedge (p \to q)$	$p \wedge q$
T	T	T	T	T
T	F	F	F	F
F	T	T	F	F
F	F	T	F	F

(Equivalent)

7. Let b stand for "Alaina likes basketball," s for "Alaina likes swimming," and g for "Alaina likes gymnastics." From the truth tables we see that (a), (c), and (d) are equivalent.

(a) $b \to (s \wedge g)$ has truth table:

b	s	g	$s \wedge g$	$b \to (s \wedge g)$
T	T	T	T	T
T	T	F	F	F
T	F	T	F	F
T	F	F	F	F
F	T	T	T	T
F	T	F	F	T
F	F	T	F	T
F	F	F	F	T

(b) $g \to (s \wedge b)$ has truth table:

b	s	g	$s \wedge b$	$g \to (s \wedge b)$
T	T	T	T	T
T	T	F	T	T
T	F	T	F	F
T	F	F	F	T
F	T	T	F	F
F	T	F	F	T
F	F	T	F	F
F	F	F	F	T

(c) $(\neg g \vee \neg s) \to \neg b$ has truth table:

b	s	g	$\neg b$	$\neg s$	$\neg g$	$\neg g \vee \neg s$	$(\neg g \vee \neg s) \to \neg b$
T	T	T	F	F	F	F	T
T	T	F	F	F	T	T	F
T	F	T	F	T	F	T	F
T	F	F	F	T	T	T	F
F	T	T	T	F	F	F	T
F	T	F	T	F	T	T	T
F	F	T	T	T	F	T	T
F	F	F	T	T	T	T	T

(d) $\neg b \vee (s \wedge g)$ has truth table:

b	s	g	$\neg b$	$s \wedge g$	$\neg b \vee (s \wedge g)$
T	T	T	F	T	T
T	T	F	F	F	F
T	F	T	F	F	F
T	F	F	F	F	F
F	T	T	T	T	T
F	T	F	T	F	T
F	F	T	T	F	T
F	F	F	T	F	T

8. (a) $(x > 0) \to (y > 0)$; (b) $(x > 0) \to (y \le 0)$; (c) $(x \le 0) \to (y > 0)$; (d) $(x \le 0) \to (y \le 0)$

10. (b) is the only statement that is not true of all elements in D. A counterexample is $x = 3$.

12. (a) $P(n)$ is "n is even," and $Q(n)$ is "$n^2 + n$ is even."

(b) $P(n)$ is "n is a multiple of 5," and $Q(n)$ is "n has ones' digit of 5."

(c) $P(n)$ is "n is prime," and $Q(n)$ is "$2^n - 1$ is prime."

14. Use $\mathbb{Z}$ to represent the set of all integers, and $\mathbb{X}$ for the set of positive real numbers. Let E_m represent, "ends in the digit m," and D_m represent, "is evenly divisible by m." For part (d), let $L(x)$ represent "$x < \sqrt{2}$," and $G(x)$ represent, "$2/x > \sqrt{2}$."

(a) $\forall n \in \mathbb{Z}, E_5(n) \to D_5(n)$

(b) $\forall n \in \mathbb{Z}, E_3(n) \to D_3(n)$

(c) $\forall n \in \mathbb{Z}, D_5(n) \to D_3(n^2 - 1)$

(d) $\forall x \in \mathbb{X}, L(x) \to G(x)$

15. Parts (a) and (d) are true. For (b), 13 is a counterexample. For (c), 15 is a counterexample. Notice that part (d) would be false if we had quantified over "non-zero real numbers" with counterexample -1.

16. Using the same sets and predicates as in Exercise 14, the negations are:

(a) $\exists n \in \mathbb{Z}, E_5(n) \wedge \neg D_5(n)$

(b) $\exists n \in \mathbb{Z}, E_3(n) \wedge \neg D_3(n)$

(c) $\exists n \in \mathbb{Z}, D_5(n) \wedge \neg D_3(n^2 - 1)$

(d) $\exists x \in \mathbb{X}, L(x) \wedge \neg G(x)$

20. (a) is false (counterexample $n = -2$); (c) is false (counterexample $x = 3\pi/2$)

22. There are many equivalent answers. We present versions consistent with the way we have written statements in the previous sections.

(a) For every positive integer n, if n is even, then $\frac{1}{n} \le 1$.

(b) For all positive integers a and b, if $a - b$ is odd, then $a^2 \neq 2b^2$.

(c) For all integers a and b, if a and b are positive, then $\frac{a}{b} \neq 1 + \frac{b}{a}$.

25. (a) If you will not get an F for the course, then you don't attend the concert.

    (c) If you will not be hungry, then you eat your breakfast.

26. (a) If you will get an F for the course, then you don't attend the concert.

    (c) If you will be hungry, then you don't eat your breakfast.

27. (a) If you attend the concert, then you will not get an F for the course.

    (c) If you eat your breakfast, you will not be hungry.

28. (a) Both are telling the truth.

	p	q	A says $\neg q \to \neg p$	B says $\neg q \to \neg p$
⋆	T	T	T	T
	T	F	F	F
	F	T	T	T
	F	F	T	T

30. (a) is equivalent to (f), (c) to (e), and (d) to (g). The pairs of negations include (a) and (d), (b) and (c).

    (a) $\forall s \in S, C(s) \to D(s)$

    (c) $\forall s \in S, C(s) \to \neg D(s)$

    (e) $\forall s \in S, C(s) \to \neg D(s)$

Section 1.6 exercises

1. No. The poor child has fallen for the *inverse fallacy*.

2. Only (d) is valid.

    (a) Let p be, "both numbers are even," and q be, "the sum is even." Then this argument is of the form $p \to q, \neg p, \therefore \neg q$, which is an example of the inverse fallacy.

    (b) Let p be, "this number is a perfect square," and q be, "the equation has a rational solution." Then this argument is of the form $p \to q, q, \therefore p$, which is an example of the converse fallacy.

    (c) Let p be, "you bought from us," and q be, "you paid too much." Then this argument is of the form $\neg p \to q, p, \therefore \neg q$, which is an example of the inverse fallacy.

    (d) Let p be, "this prime number is even," and q be, "this number is less than 5." Then this argument is of the form $p \to q, p, \therefore q$, which is an example of valid reasoning by *modus ponens*.

    (e) Let p be, "this city is large," and q be, "this city has large buildings." Then this argument is of the form $p \to q, q, \therefore p$, which is an example of the converse fallacy.

4. (a) If you read the book, then you will pass the course. (b) If you will pass the course, then you must read the book. (c) If you will pass the course, then you must read the book.

6. (a)

p	$\neg p$	$p \to \neg p$	
T	F	F	(neither)
F	T	T	

    (b)

p	q	$p \to q$	$q \to p$	$(p \to q) \vee (q \to p)$	
T	T	T	T	T	
T	F	F	T	T	(tautology)
F	T	T	F	T	
F	F	T	T	T	

9

	p	q	$p \wedge q$	$q \to \neg p$	$(p \wedge q) \vee (q \to \neg p)$	
	T	T	T	F	T	
(c)	T	F	F	T	T	(tautology)
	F	T	F	T	T	
	F	F	F	T	T	

	p	q	$p \vee \neg q$	$q \wedge \neg p$	$(p \vee \neg q) \to (q \wedge \neg p)$	
	T	T	T	F	F	
(d)	T	F	T	F	F	(neither)
	F	T	F	T	T	
	F	F	T	F	F	

8. Examples will vary.

(a) John ordered pizza and Jill ordered pizza. Conclusion: John ordered pizza.

p	q	$p \wedge q$	$(p \wedge q) \to p$
T	T	T	T
T	F	F	T
F	T	F	T
F	F	F	T

(c) Bill got a haircut. Sue went shopping. Conclusion: Bill got a haircut and Sue went shopping.

p	q	$p \wedge q$	$(p \wedge q) \to (p \wedge q)$
T	T	T	T
T	F	F	T
F	T	F	T
F	F	F	T

(e) If Keith attends the party, Meg will attend. If Meg attends the party, Nancy will attend. Conclusion: If Keith attends the party, Nancy will attend.

p	q	r	$p \to q$	$q \to r$	$(p \to q) \wedge (q \to r)$	$p \to r$	$((p \to q) \wedge (q \to r)) \to (p \to r)$
T	T	T	T	T	T	T	T
T	T	F	T	F	F	F	T
T	F	T	F	T	F	T	T
T	F	F	F	T	F	F	T
F	T	T	T	T	T	T	T
F	T	F	T	F	F	T	T
F	F	T	T	T	T	T	T
F	F	F	T	T	T	T	T

9. (a) This is not valid since the truth table shows that when p and q are both false, then the premises are all true while the conclusion of the argument is false.

p	q	$p \to q$	$q \to p$	$(p \to q) \wedge (q \to p)$	$p \wedge q$	$((p \to q) \wedge (q \to p)) \to (p \wedge q)$
T	T	T	T	T	T	T
T	F	F	T	F	F	T
F	T	T	F	F	F	T
F	F	T	T	T	F	F

(c) This is valid. The truth table is exactly like 8(e).

10. (a) N is "Newton is considered a great mathematician," L is "Leibniz work is ignored," and C is "Calculus is the centerpiece." This argument has the structure: $(\neg N \wedge \neg L) \to \neg C, N \to L \therefore C \wedge \neg L$

| | | | | $(\neg N \wedge \neg L)$ | | | $(((\neg N \wedge \neg L) \to \neg C) \wedge (N \to L))$ |
N	L	C	$\neg N \wedge \neg L$	$\to \neg C$	$N \to L$	$C \wedge \neg L$	$\to (C \wedge \neg L)$
T	T	T	F	T	T	F	F
T	T	F	F	T	T	F	F
T	F	T	F	T	F	T	T
T	F	F	F	T	F	F	T
F	T	T	F	T	T	F	F
F	T	F	F	T	T	F	F
F	F	T	T	F	T	T	T
F	F	F	T	T	T	F	F

The first row of the table shows that if N, L and C are all true, then the hypotheses of the argument are true while the conclusion is false.

(c) r is "I have a good round of golf," c is "the wind is calm," and d is "the weather is dry." This argument has the structure: $r \to (c \vee d), c \wedge d \therefore r$

r	c	d	$c \vee d$	$r \to (c \vee d)$	$c \wedge d$	$((r \to (c \vee d)) \wedge (c \wedge d)) \to r$
T	T	T	T	T	T	T
T	T	F	T	T	F	T
T	F	T	T	T	F	T
T	F	F	F	F	F	T
F	T	T	T	T	T	F
F	T	F	T	T	F	T
F	F	T	T	T	F	T
F	F	F	F	T	F	T

The fifth row of the table shows that if c and d are true but r is false, then the hypotheses of the argument are true while the conclusion is false.

11. Each argument is valid.

(a) g is "we take our gas heater," e is "we take extra blankets," and m is "we take our air mattress." This argument has the structure: $g \to e, \neg e \to \neg m, \therefore (m \vee g) \to e$

g	e	m	$g \to e$	$\neg e \to \neg m$	$m \vee g$	$(m \vee g) \to e$	$((g \to e) \wedge (\neg e \to \neg m))$ $\to ((m \vee g) \to e)$
T	T	T	T	T	T	T	T
T	T	F	T	T	T	T	T
T	F	T	F	F	T	F	T
T	F	F	F	T	T	F	T
F	T	T	T	T	T	T	T
F	T	F	T	T	F	T	T
F	F	T	T	F	T	F	T
F	F	F	T	T	F	T	T

Section 2.1 exercises

1. (a) 12; (c) -8; (e) $n^2 + 2n - 3$; (g) -12.

2. (a) A counterexample is $m = 3, n = 2$.

   (b) Seems to be true (It works for $n = 3, 9, 15, 21$, for example.)

   (c) A counterexample is $n = 9$.

   (d) A counterexample is $n = 3$.

   (e) Seems to be true (It works for $n = 1, 3, 5, 7, 9, 11, 13, 15, 17$, for example.)

   (f) Seems to be true (For none of the perfect squares $x = 1, 4, 9, 16, 25, 36, 49, 64, 81, 100, 121$, for example, is $x + 1$ a product of two even numbers.)

(g) A counterexample is $n = 6$.

(h) A counterexample is $n = 4$.

4. (a) Dear READER,

   If you choose any even number and call it n, then I can show you that $3n$ is also even. Here's how. Since n is even, we agree it can be written as $n = 2k$ for some integer k. Then if I calculate $3n$, I get $3n = 3(2k) = 6k = 2(3k)$. Now $3k$ is also an integer, so this takes the form of 2 times some integer, which is exactly what we mean by saying it is even.

   Your friend, AUTHOR

   (c) Dear READER,

   Let's call your even number n, and use our agreement on what even means to write $n = 2k$ (k being some integer). Now calculate $n + 1 = 2k + 1$. By our agreement on the meaning of odd, this shows that $n + 1$ is odd.

   Your friend, AUTHOR

   (e) Dear READER,

   This one is a little tougher. Let's start by naming your odd number n, and writing $n = 2k + 1$ (k some integer). (This is what n being odd means.) If we calculate $n^3 - n$, we get $(2k+1)^3 - (2k+1) = 8k^3 + 12k^2 + 6k + 1 - 2k - 1 = 8k^3 + 12k^2 + 4k = 4(2k^3 + 3k^2 + k)$. Since we have written $n^3 - n$ as 4 times the integer $2k^3 + 3k^2 + k$, this shows that $n^3 - n$ is divisible by 4.

   Your friend, AUTHOR

5. (a) **Proposition.** The sum of two odd integers is even.

   Proof. Let m and n be odd integers. This means there is an integer K such that $m = 2K + 1$ and there is an integer L such that $m = 2L + 1$, and so

   $$\begin{aligned} m + n &= \underline{(2K+1)} + \underline{(2L+1)} \\ &= \underline{2 \cdot (K+L+1)} \end{aligned}$$

   Since $\underline{K + L + 1}$ is an integer, this means that $m + n$ is even.

   (b) **Proposition.** If n is even, then n^2 is even.

   Proof. Let n be an even integer. This means that we can write $n = \underline{2m}$ for some integer m. This in turn means that $n^2 = 4m^2 = 2 \cdot \underline{(2m^2)}$, so n^2 is even, because $\underline{2m^2}$ is an integer.

   (c) **Proposition.** Every odd perfect square can be written in the form $4k + 1$, where k is an integer.

   Proof. Let s be an odd perfect square. So $s = n^2$ for some integer n, and n^2 is odd. By the contrapositive of the previous exericse, n is odd. Since n is odd, there is an integer L such that $n = \underline{2L + 1}$. This means

   $$\begin{aligned} s = n^2 &= \underline{(2L+1)^2} \\ &= \underline{4L^2 + 4L + 1} \end{aligned}$$

   So $s = 4(\underline{L^2 + L}) + 1$, where $\underline{L^2 + L}$ is an integer, as desired.

6. The table is shown below.

	x	y	K	L	$K + L$	$2 \cdot (K + L)$	$x + y$
(a)	18	20	9	10	19	38	38
(c)	18	18	9	9	18	36	36

9. (a) Dear READER,

   Let's agree to call your example of two odd integers x and y. Since x is odd, we know there must be an integer K that makes $x = 2K + 1$ true, and likewise we know we can write $y = 2L + 1$ for some integer L. Now let's see what happens when we multiply your two integers using some algebra:

   $$\begin{aligned} x \cdot y &= (2K + 1) \cdot (2L + 1) \\ &= 4KL + 2L + 2K + 1 \\ &= 2 \cdot (2KL + L + K) + 1 \end{aligned}$$

12

Since $2KL + L + K$ is an integer, we can see that we have written $x \cdot y$ as $2 \cdot (\text{integer}) + 1$, which shows us that $x \cdot y$ is odd.

Your friend,

AUTHOR

(c) Dear READER,

Consider an example of an even integer (which we'll call x) and an integer that is divisible by 3 (which we'll call y). Since x is even, we know there must be an integer K that makes $x = 2K$ true. Since y is divisible by 3, we know we can write $y = 3L$ for some integer L. Now let's see what happens when we multiply these two integers:

$$x \cdot y = (2K)(3L)$$
$$= 6(KL)$$

Since KL is an integer, we can see that we have written $x \cdot y$ as $6 \cdot (\text{integer})$, which is exactly what it means to say that $x \cdot y$ is divisible by 6.

Your friend,

AUTHOR

11. (a) *Proof.* Let x and y be odd integers. Then there is an integer K such that $x = 2K + 1$ and there is an integer L such that $y = 2L + 1$. It follows then that

$$x \cdot y = (2K + 1) \cdot (2L + 1)$$
$$= 4KL + 2L + 2K + 1$$
$$= 2 \cdot (2KL + L + K) + 1$$

Since $2KL + L + K$ is an integer, we can conclude that $x \cdot y$ is odd. □

(c) *Proof.* Let an even integer x and an integer y that is divisible by 3 be given. Then there is an integer K such that $x = 2K$ and there is an integer L such that $y = 3L$. It follows then that

$$x \cdot y = (2K)(3L)$$
$$= 6(KL)$$

Since KL is an integer, we can conclude that $x \cdot y$ is divisible by 6. □

12. Only the pairs in (c) and (e) are logically equivalent. The others are not:

(a) A counterexample to (i) requires that my brother and I both root for the Braves while a counterexample to (ii) requires that I do not root for the Braves while my brother does.

(c) A counterexample to each would be a student who does not do math problems every night but is good at math.

13. (a) **Proposition.** If $m = 0$ and $n = 0$, then $m^2 + n^2 = 0$.

Proof. If $m = 0$ and $n = 0$, then $m^2 + n^2 = 0^2 + 0^2 = 0$.

(c) **Proposition.** If m and n are odd integers, then $m + n$ is an even integer.

Proof. Let odd integers n and n be given. This means $n = 2K + 1$ for some integer K, and $m = 2L + 1$ for some integer L. So

$$m + n = (2K + 1) + (2L + 1)$$
$$= 2K + 2L + 2$$
$$= 2(K + L + 1)$$

which means $m + n$ can be written as twice an integer, hence $m + n$ is even.

Section 2.2 exercises

1. (a) $73 = 6 \cdot 12 + 1$
   (c) $-1234 = 15 \cdot (-83) + 11$
   (e) $1000 = 7 \cdot 142 + 6$

2. (a) $73 \bmod 6 = 1$
   (c) $-1234 \bmod 15 = 11$
   (e) $1000 \bmod 7 = 6$

3. (a) $55 \bmod 6 = 1$
   (c) $6 \bmod 55 = 6$
   (e) $\left(6(k^3 - 3k^2 - 2k + 8) + 5\right) \bmod 6 = 5$
   (g) $10(9k + 1) = 90k + 10 = 9(10k + 1) + 1$ and $(9(10k + 1) + 1) \bmod 9 = 1$

6. (a) **Proposition:** If 3 divides b and b divides c, then 3 divides c.
   Proof. Let integers b and c be given, and assume that 3 divides b and b divides c. This means that $b = 3 \cdot k$ for some integer k and $c = m \cdot b$ for some integer m. From this we can conclude that

   $$
   \begin{aligned}
   c &= m \cdot b \\
   &= m \cdot (3 \cdot k) \\
   &= 3 \cdot (mk)
   \end{aligned}
   $$

   Since $\underline{mk}$ is an integer, this establishes that 3 divides c.

   (b) **Proposition:** If n^3 is even, then so is n.
   Proof of the contrapositive. Let an odd integer n be given. Then we may choose an integer m so that $n = 2m + 1$. This means

   $$
   \begin{aligned}
   n^3 &= (2m + 1)^3 \\
   &= 8m^3 + 12m^2 + 6m + 1 \\
   &= 2(4m^3 + 6m^2 + 3m) + 1
   \end{aligned}
   $$

   Since $4m^3 + 6m^2 + 3m$ is an integer, we can see that n^3 is 1 more than twice an integer. Hence n^3 is odd, completing the proof.

   (c) **Proposition:** If 3 divides $4^{n-1} - 1$, then 3 divides $4^n - 1$
   Proof. Let the integer n be given, and assume that 3 divides $4^{n-1} - 1$. This means that $4^{n-1} - 1 = 3k$ for some integer. So $4^{n-1} = 3k + 1$, and

   $$
   \begin{aligned}
   4^n &= 4(4^{n-1}) \\
   &= 4(3k + 1) \\
   &= 12k + 4 \\
   &= 3(4k + 1) + 1
   \end{aligned}
   $$

   Therefore, $4^n - 1 = 3(\underline{4k + 1})$. Since $\underline{4k + 1}$ is an integer, this means that 3 divides $4^n - 1$.

7. (a) If a divides b and a divides c, then a divides $b + c$.
   Proof. Let integers a, b and c be given such that a divides b and a divides c. This means that $b = k \cdot a$ for some integer k and $c = l \cdot a$ for some integer l, so $b + c = k \cdot a + l \cdot a = (k + l) \cdot a$. Since $k + l$ is an integer, this means that a divides $b + c$.

   (c) If a divides b and c divides d, then ac divides bd.
   Proof. Let integers a, b, c and d be given such that a divides b and c divides d. This means that $b = k \cdot a$ for some integer k and $d = l \cdot c$ for some integer l, so $b \cdot d = (k \cdot a) \cdot (l \cdot c) = (kl) \cdot (ac)$. Since kl is an integer, this means that ac divides bd.

(e) If 9 divides $10^{n-1} - 1$, then 9 divides $10^n - 1$.

Proof. Let an integer n be given such that 9 divides $10^{n-1} - 1$. This means that $10^{n-1} - 1 = 9k$ for some integer k. We can show algebraically that

$$
\begin{aligned}
10^n - 1 &= 10 \cdot (10^{n-1} - 1) + 9 \\
&= 10 \cdot (9k) + 9 \\
&= 9 \cdot (10k + 1)
\end{aligned}
$$

Since $10k + 1$ is an integer, this means that 9 divides $10^n - 1$.

8. **Proposition:** The sum of two rational numbers is a rational number. (That is, if a and b are rational numbers, then $a + b$ is a rational number.)

Proof. Let rational numbers a and b be given. Since <u>a is rational</u>, we know that $a = \frac{x}{y}$ for some integers x and y with $y \neq 0$. Likewise, since <u>b is rational</u>, we know that $b = \frac{z}{w}$ for some integers z and w with $w \neq 0$. From the rules for adding fractions, we know that

$$
a + b = \frac{xw + yz}{yw}
$$

We know that $xw + yz$ and yw are both integers, and $yw \neq 0$ because <u>neither y nor w is 0</u>. Hence, we know that $a + b$ is rational, by the definition of "rational." □

9. **Proposition:** The difference between two rational numbers is a rational number. (That is, if a and b are rational numbers, then $a - b$ is a rational number.)

Proof. Let rational numbers a and b be given. Since a is rational, we know that $a = \frac{x}{y}$ for some integers x and y with $y \neq 0$. Likewise, since b is rational, we know that $b = \frac{z}{w}$ for some integers z and w with $w \neq 0$. From the rules for subtracting fractions, we know that

$$
a + b = \frac{xw - yz}{yw}
$$

We know that $xw - yz$ and yw are both integers, and $yw \neq 0$ because neither y nor w is 0. Hence, we know that $a - b$ is rational, by the definition of "rational." □

11. **Proposition.** The average of two rational numbers is a rational number.

Proof. Let x and y be rational numbers. Then there are integers a and b (with $b \neq 0$) such that $x = \frac{a}{b}$, and there are integers c and d (with $d \neq 0$) such that $y = \frac{c}{d}$. So

$$
\frac{x + y}{2} = \frac{\frac{a}{b} + \frac{c}{d}}{2} = \frac{ad + bc}{2bd}.
$$

Since $ad + bc$ and $2bd$ are integers (with $2bd \neq 0$), this shows that $\frac{x+y}{2}$ is a rational number. □

13. (a) For the sequence $a_n = a_{n-1} + 2n \ldots$

    i. *Proof.* Let n be given such that a_{n-1} is even. This means that $a_{n-1} = 2k$ for some integer k. Therefore, $a_n = a_{n-1} + 2n = 2k + 2n = 2(k + n)$, which is even.

    ii. $10, 14, 20, 28, 38, \ldots$. Every term is even.

    iii. $7, 11, 17, 25, 35, \ldots$. Since the statement "every even term is followed by an even term" has no counterexample, we must say that this statement is true for all terms.

14. (a) *Proof.* Considering the contrapositive statement, let an integer n be given which is not divisible by 3. By the Division Theorem, when any integer is divided by 3 it leaves a remainder of 0, 1 or 2. That is, one of the following cases must be true:

    • **Case 1:** It might be that $n = 3q$ for some integer q. However, for this particular integer n, we know this case does not happen, because the given n is not divisible by 3.

- **Case 2:** It might be that $n = 3q + 1$ for some integer q. In this case,

$$n^2 = (3k+1)^2 = 9k^2 + 6k + 1$$
$$= 3 \cdot (3k^2 + 2k) + 1$$

- **Case 3:** It might be that $n = 3q + 2$ for some integer q. In this case,

$$n^2 = (3k+2)^2 = 9k^2 + 12k + 4$$
$$= 3 \cdot (3k^2 + 4k + 1) + 1$$

Thus, in every case that satisfies the hyphothesis, we see that n^2 is not divisible by 3, completing the proof.

16. *Proof.* Let an integer n which is not divisible by 3 be given. By the Division Theorem, when any integer is divided by 3 it leaves a remainder of 0, 1 or 2. That is, one of the following cases must be true:

- **Case 1:** It might be that $n = 3q$ for some integer q. However, for this particular integer n, we know this case does not happen because the given n is not divisible by 3.
- **Case 2:** It might be that $n = 3q + 1$ for some integer q. In this case,

$$n^2 + 2 = (3k+1)^2 + 2$$
$$= \left(9k^2 + 6k + 1\right) + 2$$
$$= 3 \cdot \left(3k^2 + 2k + 1\right)$$

- **Case 3:** It might be that $n = 3q + 2$ for some integer q. In this case,

$$n^2 + 2 = (3k+2)^2 + 2$$
$$= \left(9k^2 + 12k + 4\right) + 2$$
$$= 3 \cdot \left(3k^2 + 4k + 2\right)$$

Thus, in every case that satisfies the hyphothesis, we see that $n^2 + 2$ is divisible by 3, completing the proof.

18. *Proof.* We first show that $n^3 + 2n$ is divisible by 3. We write $n^3 + 2n = n(n^2 + 2)$. Now either n is divisible by 3, or it is not.

Case 1. If n is divisible by 3, say $n = 3k$ for some integer k, then $n^3 + 2n = n(n^2 + 2) = 3k(n^2 + 2) = 3(kn^2 + 2k)$, so $n^3 + 2n$ is divisible by 3.

Case 2. If n is not divisible by 3, then by Exercise 16, we know that $n^2 + 2$ is divisible by 3, say $n^2 + 2 = 3m$ for some integer m. So $n^3 + 2n = n(n^2 + 2) = n(3m) = 3(nm)$, so $n^3 + 2n$ is divisible by 3.

Thus in either case, $n^3 + 2n$ is divisible by 3. But in Exercise 17 we also showed that $n^3 + 2n$ is divisible by 4, so by Exercise 15, we conclude that $n^3 + 2n$ is divisible by 12.

20. The sum of any three consecutive perfect cubes is divisible by 9.

Proof. Call the three consecutive perfect cubes, $(n-1)^3$, n^3 and $(n+1)^3$. Then

$$(n-1)^3 + n^3 + (n+1)^3 = 3n^3 + 6n$$
$$= 3n(n^2 + 2)$$

Now n is either divisible by 3 or it is not.

- **Case 1:** If n is divisible by 3, then $3n$ is divisible by 9, and hence $3n(n^2 + 2)$ is divisible by 9.
- **Case 2:** If n is not divisible by 3, by Exercise 16 we know that $n^2 + 2$ is divisible by 3, and hence $3\left(n^2 + 2\right)$ is divisible by 9. It follows then that $3n(n^2 + 2)$ is divisible by 9.

In either case, $3n(n^2 + 2)$ is divisible by 9.

22. For all integers n, $n^5 - n$ is divisible by 5.

Proof. Let an integer n be given. By the Division Theorem, when n is divided by 5, it leaves a remainder of 0, 1, 2, 3, or 4.

- **Case 1:** It might be that $n = 5q$ for some integer q. In this case,

$$n^5 - n = 3125q^5 - 5q$$
$$= 5 \cdot (625q^5 - q)$$

- **Case 2:** It might be that $n = 5q + 1$ for some integer q. In this case,

$$n^5 - n = (5q + 1)^5 - (5q + 1)$$
$$= 3125q^5 + 3125q^4 + 1250q^3 + 250q^2 + 20q$$
$$= 5 \cdot (625q^5 + 625q^4 + 250q^3 + 50q^2 + 4q)$$

- **Case 3:** It might be that $n = 5q + 2$ for some integer q. In this case,

$$n^5 - n = (5q + 2)^5 - (5q + 2)$$
$$= 3125q^5 + 6250q^4 + 5000q^3 + 2000q^2 + 395q + 30$$
$$= 5 \cdot (625q^5 + 1250q^4 + 1000q^3 + 400q^2 + 79q + 6)$$

- **Case 4:** It might be that $n = 5q + 3$ for some integer q. In this case,

$$n^5 - n = (5q + 3)^5 - (5q + 3)$$
$$= 3125q^5 + 9375q^4 + 11\,250q^3 + 6750q^2 + 2020q + 240$$
$$= 5 \cdot (625q^5 + 1875q^4 + 2250q^3 + 1350q^2 + 404q + 48)$$

- **Case 5:** It might be that $n = 5q + 4$ for some integer q. In this case,

$$n^5 - n = (5q + 4)^5 - (5q + 4)$$
$$= 3125q^5 + 12500q^4 + 20000q^3 + 16000q^2 + 6395q + 1020$$
$$= 5 \cdot (625q^5 + 2500q^4 + 4000q^3 + 3200q^2 + 1279q + 204)$$

Hence, in every possible case, $n^5 - n$ is divisible by 5.

24. *Proof.* Let A, B, C and D be given such that $5A + B = 5C + D$ with $0 \leq B \leq 4$ and $0 \leq D \leq 4$. Since this is the same as $5(A - C) = D - B$, $D - B$ is divisible by 5. But

$$-4 \leq D - B \leq 4$$

and 0 is the only number in this interval that is divisible by 5, so we can conclude that $B = D$. From this it follows that $5(A - C) = 0$ because $5(A - C) = D - B = 0$, from which it follows that $A = C$ as well.

26. (a) The contrapositive of the statement, "If n^2 is even, then so is n" is the statement, "If n is odd, then n^2 is odd."

Proof. Let n be a given odd integer. Since n is odd, we can write $n = 2k + 1$ where k is an integer. Then $n^2 = (2k + 1)^2 = 4k^2 + 4k + 1 = 2(2k^2 + 2k) + 1$. This shows that n^2 is odd.

(b) This was proven in Proposition 7.

(c) *Proof.* Let integers a, b, and c, with $a^2 + b^2 = c^2$ and c being even, be given. Since c is even, we can write $c = 2k$ for some integer k. There are two possibilities for a — either it is even or it is odd. We consider both cases:

Case 1. a is even. Then $a = 2m$ for some integer m. In this case, the equation $a^2 + b^2 = c^2$ can be rewritten $b^2 = c^2 - a^2 = 4k^2 - 4m^2 = 2(2k^2 - 2m^2)$, from which it follows that b^2 is even, and thus b is even by part (a).

Case 2. a is odd. Then $a = 2m + 1$ for some integer m. In this case, the equation $a^2 + b^2 = c^2$ can be rewritten $b^2 = c^2 - a^2 = 4k^2 - (4m^2 + 4m + 1) = 4k^2 - 4m^2 - 4m - 1 = 4(k^2 - m^2 - m - 1) + 3$. By part (b), it is impossible for a perfect square to be written in the form $4M + 3$ where M is an integer.

Since Case 2 is impossible, it must be the case that a is even, from which it follows that b is also even.

28. *Proof.* Let n be a perfect square integer. This means that $n = m^2$ for some integer m. The Division Theorem tells us that when m is divided by 3, it must leave a remainder of 0, 1 or 2.

Case 1. If $m = 3q$, then $n = (3q)^2 = 9q^2$, which is not of the form $3k + 2$.

Case 2. If $m = 3q + 1$, then $n = (3q + 1)^2 = 9q^2 + 6q + 1 = 3(3q^2 + 2q) + 1$, which is not of the form $3k + 2$.

Case 3. If $m = 3q + 2$, then $n = (3q + 2)^2 = 9q^2 + 12q + 4 = 3(3q^2 + 4q + 1) + 1$, which is not of the form $3k + 2$.

In no case is n of the form $3k + 2$.

30. *Proof.* Let an integer c be given. By the Division Theorem, dividing c by 4 leaves a remainder of 0, 1, 2, or 3.

- **Case 1:** It might be that $c = 4q$ for some integer q. In this case, $c + 2 = 4q + 2$, which is an impossible form for a perfect square by Proposition 7.

- **Case 2:** It might be that $c = 4q + 1$ for some integer q. In this case, $c + 2 = 4q + 3$, which is an impossible form for a perfect square by Proposition 7.

- **Case 3:** It might be that $c = 4q + 2$ for some integer q. In this case, $2c + 2 = 8q + 4 + 2 = 4(2q + 1) + 2$, which is an impossible form for a perfect square by Proposition 7.

- **Case 4:** It might be that $c = 4q + 3$ for some integer q. In this case, $7c + 2 = 28q + 21 + 2 = 4(7q + 5) + 3$, which is an impossible form for a perfect square by Proposition 7.

In every case, at least one of the values $c + 2$, $2c + 2$, or $7c + 2$ is not a perfect square.

32. *Outline of proof.* We should first establish that every perfect square is of the form $8k$, $8k + 1$, or $8k + 4$. This can be done by extending the reasoning in Proposition 7. We now consider three perfect squares a^2, b^2 and c^2, and list all the possibilities for their sum, in the form $8M + ??$. The only form not possible is $8M + 7$.

Section 2.3 exercises

1. (a) $P(1)$ is "2 is prime," $P(2)$ is "5 is prime," and $P(12)$ is "145 is prime." Both $P(1)$ and $P(2)$ are true. $P(m - 1)$ is "$(m - 1)^2 + 1$ is prime."

(b) $L(1)$ is "$1 < 2$," $L(2)$ is "$4 < 4$," $L(3)$ is "$9 < 8$," $L(4)$ is "$16 < 16$," $L(5)$ is "$25 < 32$," and $L(6)$ is "$36 < 64$." Of these, $L(1)$, $L(5)$ and $L(6)$ are true. $L(m - 1)$ is "$(m - 1)^2 < 2^{m-1}$."

(c) $S(1)$ is "$1^2 = \frac{1(2)(3)}{6}$," $S(2)$ is "$1^2 + 2^2 = \frac{2(3)(5)}{6}$," $S(3)$ is "$1^2 + 2^2 + 3^2 = \frac{3(4)(7)}{6}$," $L(4)$ is "$1^2 + 2^2 + 3^2 + 4^2 = \frac{4(5)(9)}{6}$," $L(5)$ is "$1^2 + 2^2 + 3^2 + 4^2 + 5^2 = \frac{5(6)(11)}{6}$," and $L(6)$ is "$1^2 + 2^2 + 3^2 + 4^2 + 5^2 + 6^2 = \frac{6(7)(13)}{6}$." Each of these is true. $S(m - 1)$ is "$\sum_{i=1}^{m-1} i^2 = \frac{(m-1)(m)(2m-1)}{6}$."

3. (a) *Proof.* Let $P(n)$ be the statement, "$a_n = 4n - 3$." We verify statements $P(1), P(2), P(3), P(4)$ in table form:

n	a_n (via recursive formula)	$4n-3$	equal?
1	1	$4 \cdot 1 - 3 = 1$	yes
2	$1 + 4 = 5$	$4 \cdot 2 - 3 = 5$	yes
3	$5 + 4 = 9$	$4 \cdot 3 - 3 = 9$	yes
4	$9 + 4 = 13$	$4 \cdot 4 - 3 = 13$	yes

Now let the integer $m \geq 2$ be given such that $P(1), P(2), \ldots, P(m-1)$ have already been checked to be true. In particular, the last statement we checked was "$a_{m-1} = 4(m-1) - 3$." It now follows that

$$
\begin{aligned}
a_m &= a_{m-1} + 4 \text{ (by the recurrence relation for } a) \\
&= (4(m-1) - 3) + 4 \text{ (by } P(m-1)) \\
&= 4m - 4 - 3 + 4 \\
&= 4m - 3
\end{aligned}
$$

Thus we have shown that $a_m = 4m - 3$, which is precisely statement $P(m)$.

(c) *Proof.* Let $P(n)$ be the statement, "$a_n = \frac{n(n+1)(2n+1)}{2}$." We verify statements $P(1), P(2), P(3), P(4)$ in table form:

n	a_n (via recursive formula)	$\frac{n(n+1)(2n+1)}{2}$	equal?
1	1	$\frac{1(2)(3)}{6} = 1$	yes
2	$1 + 2^2 = 5$	$\frac{2(3)(5)}{6} = 5$	yes
3	$5 + 3^2 = 14$	$\frac{3(4)(7)}{6} = 14$	yes
4	$14 + 4^2 = 30$	$\frac{4(5)(9)}{6} = 30$	yes
...	...	...	...
$m-1$	$a_{m-2} + (m-1)^2 = \frac{(m-1)(m)(2m-1)}{6}$	$\frac{(m-1)(m)(2m-1)}{6}$	yes
m	$a_{m-1} + m^2$	$\frac{m(m+1)(2m+1)}{6}$	???

Now let the integer $m \geq 2$ be given such that $P(1), P(2), \ldots, P(m-1)$ have already been checked to be true. In particular, the last statement we checked was "$a_{m-1} = \frac{(m-1)(m)(2m-1)}{2}$." It now follows that

$$
\begin{aligned}
a_m &= a_{m-1} + m^2 \text{ (by the recurrence relation for } a) \\
&= \frac{(m-1)(m)(2m-1)}{6} + m^2 \text{ (by } P(m-1)) \\
&= \frac{(m-1)(m)(2m-1) + 6m^2}{6} \\
&= \frac{m\left[(m-1)(2m-1) + 6m\right]}{6} \\
&= \frac{m\left(2m^2 - 3m + 1 + 6m\right)}{6} = \frac{m\left(2m^2 + 3m + 1\right)}{6} \\
&= \frac{m(m+1)(2m+1)}{6}
\end{aligned}
$$

Thus we have shown that $a_m = \frac{m(m+1)(2m+1)}{6}$, which is precisely statement $P(m)$.

(d) *Proof.* Let $P(n)$ be the statement, "$a_n = 2^n - 1$." We verify statements $P(1), P(2), P(3), P(4)$ in table form:

n	a_n (via recursive formula)	$2^n - 1$	equal?
1	1	$2^1 - 1 = 1$	yes
2	$2 \cdot 1 + 1 = 3$	$2^2 - 1 = 3$	yes
3	$2 \cdot 3 + 1 = 7$	$2^3 - 1 = 7$	yes
4	$2 \cdot 7 + 1 = 15$	$2^4 - 1 = 15$	yes

Now let the integer $m \geq 2$ be given such that $P(1), P(2), \ldots, P(m-1)$ have already been checked to be true. In particular, the last statement we checked was "$a_{m-1} = 2^{m-1} - 1$." It now follows that

$$
\begin{aligned}
a_m &= 2a_{m-1} + 1 \text{ (by the recurrence relation for } a) \\
&= 2\left(2^{m-1} - 1\right) + 1 \text{ (by } P(m-1)) \\
&= 2^m - 2 + 1 = 2^m - 1
\end{aligned}
$$

Thus we have shown that $a_m = 2^m - 1$, which is precisely statement $P(m)$.

6. Without sigma notation, $S(n)$ can be written,

$$
\frac{1(2)}{2} + \frac{2(3)}{2} + \frac{3(4)}{2} + \cdots + \frac{n(n+1)}{2} = \frac{n(n+1)(n+2)}{6}
$$

and $S(m-1)$ can be written,

$$
\frac{1(2)}{2} + \frac{2(3)}{2} + \frac{3(4)}{2} + \cdots + \frac{(m-1)(m)}{2} = \frac{(m-1)(m)(m+1)}{6}
$$

Statement $S(1)$ is "$\frac{1(2)}{2} = \frac{1(2)(3)}{6}$," $S(2)$ is "$\frac{1(2)}{2} + \frac{2(3)}{2} = \frac{2(3)(4)}{6}$," $S(3)$ is "$\frac{1(2)}{2} + \frac{2(3)}{2} + \frac{3(4)}{2} = \frac{3(4)(5)}{6}$," and $S(4)$ is "$\frac{1(2)}{2} + \frac{2(3)}{2} + \frac{3(4)}{2} + \frac{4(5)}{2} = \frac{4(5)(6)}{6}$." Each of these is true.

8. (a) *Proof.* Let $P(n)$ be the statement, "$\sum_{i=1}^{n}(2i-1) = n^2$." We verify each of $P(1), P(2), P(3), P(4)$ in table form:

n	summation	n^2	equal?
1	1	$1^2 = 1$	yes
2	$1 + 3 = 4$	$2^2 = 4$	yes
3	$1 + 3 + 5 = 9$	$3^2 = 9$	yes
4	$1 + 3 + 5 + 7 = 16$	$4^2 = 16$	yes

Now let the integer $m \geq 2$ be given such that $P(1), P(2), \ldots, P(m-1)$ have already been checked to be true. In particular, the last statement we checked was $P(m-1)$, which said, "$\sum_{i=1}^{m-1}(2i-1) = (m-1)^2$." It now follows that

$$
\begin{aligned}
\sum_{i=1}^{m}(2i-1) &= \left(\sum_{i=1}^{m-1}(2i-1)\right) + (2m-1) \\
&= (m-1)^2 + (2m-1) \text{ (by } P(m-1)) \\
&= m^2 - 2m + 1 + (2m-1) \\
&= m^2
\end{aligned}
$$

Thus we have shown that $\sum_{i=1}^{m}(2i-1) = m^2$, which is precisely statement $P(m)$.

(c) *Proof.* Let $P(n)$ be the statement, "$\sum_{i=1}^{n}(2^i - 1) = 2^{n+1} - n - 2$." We verify each of $P(1), P(2), P(3), P(4)$ in table form:

n	summation	$2^{n+1} - n - 2$	equal?
1	1	$2^2 - 1 - 2 = 1$	yes
2	$1 + 3 = 4$	$2^3 - 2 - 2 = 4$	yes
3	$1 + 3 + 7 = 11$	$2^4 - 3 - 2 = 11$	yes
4	$1 + 3 + 7 + 15 = 26$	$2^5 - 4 - 2 = 26$	yes

Now let the integer $m \geq 2$ be given such that $P(1), P(2), \ldots, P(m-1)$ have already been checked to be true. In particular, the last statement we checked was $P(m-1)$, which said, "$\sum_{i=1}^{m-1}(2^i - 1) =$

$2^m - (m-1) - 2$." It now follows that

$$\sum_{i=1}^{m}(2^i - 1) = \left(\sum_{i=1}^{m-1}(2^i - 1)\right) + (2^m - 1)$$

$$= 2^m - (m-1) - 2 + (2^m - 1) \text{ (by } P(m-1))$$

$$= 2 \cdot 2^m - m + 1 - 2 - 1$$

$$= 2^{m+1} - m - 2$$

Thus we have shown that $\sum_{i=1}^{m}(2^i - 1) = 2^{m+1} - m - 2$, which is precisely statement $P(m)$.

(e) *Proof.* Let $P(n)$ be the statement, "$\sum_{i=1}^{n}\frac{1}{2^i} = 1 - \frac{1}{2^n}$." We verify each of $P(1), P(2), P(3), P(4)$ in table form:

n	summation	$1 - \frac{1}{2^n}$	equal?
1	$\frac{1}{2}$	$1 - \frac{1}{2} = \frac{1}{2}$	yes
2	$\frac{1}{2} + \frac{1}{4} = \frac{3}{4}$	$1 - \frac{1}{4} = \frac{3}{4}$	yes
3	$\frac{1}{2} + \frac{1}{4} + \frac{1}{8} = \frac{7}{8}$	$1 - \frac{1}{8} = \frac{7}{8}$	yes
4	$\frac{1}{2} + \frac{1}{4} + \frac{1}{8} + \frac{1}{16} = \frac{15}{16}$	$1 - \frac{1}{16} = \frac{15}{16}$	yes

Now let the integer $m \geq 2$ be given such that $P(1), P(2), \ldots, P(m-1)$ have already been checked to be true. In particular, the last statement we checked was $P(m-1)$, which said, "$\sum_{i=1}^{m-1}\frac{1}{2^i} = 1 - \frac{1}{2^{m-1}}$." It now follows that

$$\sum_{i=1}^{m}\frac{1}{2^i} = \left(\sum_{i=1}^{m-1}\frac{1}{2^i}\right) + \frac{1}{2^m}$$

$$= 1 - \frac{1}{2^{m-1}} + \frac{1}{2^m} \text{ (by } P(m-1))$$

$$= 1 - \frac{2}{2^m} + \frac{1}{2^m} = 1 - \frac{1}{2^m}$$

Thus we have shown that $\sum_{i=1}^{m}\frac{1}{2^i} = 1 - \frac{1}{2^m}$, which is precisely statement $P(m)$.

9. (a) *Proof.* Let $P(n)$ be the statement, "$\frac{(1)(2)}{2} + \frac{(2)(3)}{2} + \cdots + \frac{(n)(n+1)}{2} = \frac{n(n+1)(n+2)}{6}$". Since $P(1)$ is the statement "$\frac{(1)(2)}{2} = \frac{(1)(2)(3)}{6}$", we know that it is true. Now let $m \geq 2$ be given such that we have already checked $P(1), P(2), \ldots, P(m-1)$ to be true. Since

$$\frac{(1)(2)}{2} + \frac{(2)(3)}{2} + \cdots + \frac{(m-1)(m)}{2} + \frac{(m)(m+1)}{2}$$

$$= \left[\frac{(1)(2)}{2} + \frac{(2)(3)}{2} + \cdots + \frac{(m-1)(m)}{2}\right] + \frac{(m)(m+1)}{2}$$

$$= \frac{(m-1)(m)(m+1)}{6} + \frac{(m)(m+1)}{2} \quad \text{(since } P(m-1) \text{ is true)}$$

$$= \frac{(m)(m+1)(m-1)}{6} + \frac{(m)(m+1)(3)}{6}$$

$$= \frac{(m)(m+1)(m+2)}{6}$$

we see that $P(m)$ is also true.

10. (a) can be written

$$\sum_{i=1}^{n}\frac{i(i+1)}{2} = \frac{n(n+1)(n+2)}{6}$$

11. **Claim.** For all $n \geq 1$, $\sum_{i=1}^{n}(i)(2^i) = (n-1)2^{n+1} + 2$

21

Proof. Let $S(n)$ represent the statement, "$\sum_{i=1}^{n}(i)(2^i) = (n-1)2^{n+1}+2$." Then $S(1)$ is "$(1)(2^1) = (1-1)2^2+2$," which is true. Now let $m \geq 2$ be given such that $P(1), P(2), \ldots, P(m-1)$ have all been checked to be true. In particular, $P(m-1)$ is "$\sum_{i=1}^{m-1}(i)(2^i) = (m-2)2^m+2$." So

$$\sum_{i=1}^{m}(i)(2^i) = \left(\sum_{i=1}^{m-1}(i)(2^i)\right) + (m)(2^m)$$
$$= ((m-2)2^m+2) + (m)(2^m) \text{ by } P(m-1)$$
$$= 2 \cdot m \cdot 2^m - 2 \cdot 2^m + 2$$
$$= m \cdot 2^{m+1} - 2^{m+1} + 2$$
$$= (m-1) \cdot 2^{m+1} + 2$$

which verifies that statement $P(m)$ is true.

Section 2.4 exercises

1.   (a) $a_n = a_{n-1} + 2 \cdot 3^{n-1}$ with $a_1 = 2$.

    (c) $c_n = c_{n-1} + \frac{1}{n(n+1)}$ with $c_1 = \frac{1}{2}$

2.   (a) *Proof.* It is easy to see that $a_1 = 3^1 - 1$ since the definition above gives us that $a_1 = 2$. Let $m \geq 2$ be given such that the closed formula has been checked to work for $a_1, a_2, \ldots, a_{m-1}$. In particular, it has been checked that $a_{m-1} = 3^{m-1} - 1$. Now

$$a_m = a_{m-1} + 2 \cdot 3^{m-1} \text{ by the recurrence for } a$$
$$= \left(3^{m-1} - 1\right) + 2 \cdot 3^{m-1}$$
$$= 3 \cdot 3^{m-1} - 1 = 3^m - 1$$

    So $a_m = 3^m - 1$, completing the induction.

  (c) *Proof.* It is easy to see that $c_1 = \frac{1}{1+1}$ since the definition above gives us that $c_1 = \frac{1}{2}$. Let $m \geq 2$ be given such that the given closed formula has been checked to work for $c_1, c_2, \ldots, c_{m-1}$. In particular, it has been checked that $c_{m-1} = \frac{m-1}{m}$. Now

$$c_m = c_{m-1} + \frac{1}{m(m+1)} \text{ by the recurrence for } c$$
$$= \left(\frac{m-1}{m}\right) + \frac{1}{m(m+1)}$$
$$= \frac{(m-1)(m+1)}{m(m+1)} + \frac{1}{m(m+1)}$$
$$= \frac{m^2}{m(m+1)} = \frac{m}{m+1}$$

    So $c_m = \frac{m}{m+1}$, completing the induction.

3. (a) Let $g_n = n^3 + 2n$. Note that $g_{n-1} = (n-1)^3 + 2(n-1) = n^3 - 3n^2 + 5n - 3 = (n^3 + 2n) - 3n^2 + 3n - 3$, so $g_{n-1} = g_n - 3(n^2 - n + 1)$, or equivalently, $g_n = g_{n-1} + 3(n^2 - n + 1)$.

4. The proofs are similar to the previous exercise.

  (a) *Proof.* Let $P(n)$ be the statement, "$n^2 - n$ is divisible by 2." $P(1)$ says, "$1^2 - 1$ is divisible by 2," which is true. Let $m \geq 2$ be given such that $P(1), P(2), \ldots, P(m-1)$ have all been checked to be true. In particular, $P(m-1)$ states, "$(m-1)^2 - (m-1)$ is divisible by 2," so there is an integer K

22

such that $(m-1)^2 - (m-1) = 2K$. Algebraically, this is the same thing as writing $m^2 - 3m + 2 = 2K$. Now consider the next statement $P(m)$:

$$m^2 - m = (m^2 - 3m + 2) + (2m - 2)$$
$$= 2K + (2m - 2)$$
$$= 2(K + m - 1)$$

Hence, $m^2 - m$ is divisible by 2, verifying that statement $P(m)$ is true.

5. *Proof.* Let $P(n)$ be the statement, "$10^n - 1$ is divisible by 9." Then $P(0)$ is the statement, "$10^0 - 1$ is divisible by 9," that is, "0 is divisible by 9." This is true since $0 = 0 \cdot 9$. Now let $m \geq 1$ be given such that $P(0), P(1), \ldots, P(m-1)$ have all been verified. In particular, we know that $10^{m-1} - 1$ is divisible by 9, say $10^{m-1} - 1 = 9k$ for some integer k. Using algebra, we write:

$$
\begin{aligned}
10^{m-1} - 1 &= 9k \\
10^{m-1} &= 9k + 1 \\
10 \cdot 10^{m-1} &= 10(9k + 1) \\
10^m &= 90k + 10 \\
10^m - 1 &= 90k + 10 - 1 \\
&= 90k + 9 \\
&= 9(10k + 1)
\end{aligned}
$$

Since $10k + 1$ is an integer, this shows that $P(m)$ is true, and the result follows by induction.

8. The Fibonacci numbers are defined by $F_n = F_{n-1} + F_{n-2}$ with $F_1 = F_2 = 1$.

   (a) *Proof.* It is easy to check that $F_1 < 2^1$ and $F_2 < 2^2$ since these values ($F_1 = F_2 = 1$) are given in the definition of the Fibonacci numbers. Let $m \geq 3$ be given such that the inequality "$F_n < 2^n$" has been checked for the terms $F_1, F_2, \ldots, F_{m-1}$. In particular, we know that $F_{m-1} < 2^{m-1}$ and $F_{m-2} < 2^{m-2}$. Now

$$
\begin{aligned}
F_m &= F_{m-1} + F_{m-2} \\
&< 2^{m-1} + 2^{m-2} \\
&< 2^{m-1} + 2^{m-1} \\
&= 2^m
\end{aligned}
$$

   Hence, $F_m < 2^m$, completing the induction.

   (b) *Proof.* Let $P(n)$ be the statement, "$F_2 + F_4 + \cdots + F_{2n} = F_{2n+1} - 1$." It is easy to check that $P(1)$ (which says "$F_2 = F_3 - 1$") is true since $F_2 = 1$ and $F_3 = F_2 + F_1 = 2$. Let $m \geq 2$ be given such that $P(1), P(2), \ldots, P(m-1)$ have already been checked to be true. In particular, $P(m-1)$ says, "$F_2 + F_4 + \cdots + F_{2m-2} = F_{2m-1} - 1$." Now

$$
\begin{aligned}
F_2 + F_4 + \cdots + F_{2m} &= (F_2 + F_4 + \cdots + F_{2m-2}) + F_{2m} \\
&= (F_{2m-1} - 1) + F_{2m} \\
&= F_{2m+1} - 1
\end{aligned}
$$

   This verifies that $P(m)$ is true, completing the induction.

9. The Fibonacci numbers are defined by $F_n = F_{n-1} + F_{n-2}$ with $F_1 = F_2 = 1$.

   (a) **Claim:** For all $n \geq 1$, F_{4n} is divisible by 3.
   Proof by induction on n. We can compute $F_4 = F_3 + F_2 = 2 + 1 = 3$ to see that the first statement (which is "F_4 is divisible by 3") is true. Let $m \geq 2$ be given such that $F_4, F_8, F_{12}, \ldots, F_{4(m-1)}$ have

all been checked to be divisible by $\dot{3}$. In particular, since we know that F_{4m-4} is divisible by 3, we know there is an integer K such that $F_{4m-4} = 3K$. Now using the recurrence for the Fibonacci sequence, we see that

$$\begin{aligned}
F_{4m} &= F_{4m-1} + F_{4m-2} \\
&= (F_{4m-2} + F_{4m-3}) + (F_{4m-3} + F_{4m-4}) \\
&= (F_{4m-3} + F_{4m-4}) + 2F_{4m-3} + F_{4m-4} \\
&= 3F_{4m-3} + 2F_{4m-4} = 3F_{4m-3} + 2(3K) \\
&= 3(F_{4m-3} + 2K)
\end{aligned}$$

From this, it follows that F_{4m} is divisible by 3.

10. (a) *Proof.* Let $P(n)$ be the statement, "In the Josephus game with 2^n people, Joe's friend should stand in position $2^{n-1} + 1$." Since $P(1)$ is the statement "In the Josephus game with 2 people, Joe's friend should stand in position 2", we know that it is true. Now let $m \geq 1$ be given, and assume that we already know $P(1), P(2), \ldots, P(m-1)$ to all be true. In the play of the game with 2^m people, the order of elimination is $2, 4, 6, \ldots, 2^m$ leaving the 2^{m-1} people labeled $1, 3, 5, 7, \ldots, 2^m - 1$ for the next round of elimination. Since $P(m-1)$ is true, this game will eliminate the person in position $2^{m-2} + 1$ next to last. Because of their labels, we determine that the last remaining person was originally labeled the $(2^{m-2} + 1)^{th}$ odd number which is $2(2^{m-2} + 1) - 1 = 2^{m-1} + 1$. This establishes the truth of $P(m)$.

12. *Proof.* Let $P(n)$ be the statement, "the product of n odd integers is an odd integer." The first statement $P(1)$ states, "the product of 1 odd integer is an odd integer," which is strange to say, but certainly true. We have proved $P(2)$ before (in Exercise 9(a) of Section 2.1), but here is a quick review of that proof: Let x and y be odd integers, and write $x = 2K + 1$ and $y = 2L + 1$ for integers K and L. Then $xy = (2K+1)(2L+1) = 2(2KL + K + L) + 1$. Since $2KL + K + L$ is an integer, this show that xy is odd.

Now let $m \geq 3$ be given such that $P(2), P(3), \ldots, P(m-1)$ have all been verified, and let $a_1, a_2, \ldots, a_m$ be m even integers. We must show that $a_1 a_2 \ldots a_m$ is an odd integer. Let p indicate the product $a_1 a_2 \ldots a_{m-1}$, so that $a_1 a_2 \ldots a_m = p \cdot a_m$ By $P(m-1)$, we know that p is odd. Hence by $P(2)$ we conclude that $p \cdot a_m$ is odd, and this establishes $P(m)$. The result follows by induction.

14. (a) *Proof.* Let $P(n)$ be the statement, "there exist integers q and r such that $n = 3 \cdot q + r$ and $0 \leq r \leq 2$." If n is less than 3, this statement is clearly true (just use $q = 0$ and $r = n$). This establishes $P(0), P(1)$, and $P(2)$. Now let $m \geq 3$ be given such that $P(0), P(1), P(2) \ldots, P(m-1)$ have all been verified. By $P(m-3)$ we know we can write $m - 3 = 3 \cdot q + r$ with $0 \leq r \leq 2$. Adding 3 to both sides, we obtain $m = 3 \cdot q + r + 3 = 3 \cdot (q+1) + r$. Since $q + 1$ and r are integers and r still satisfies $0 \leq r \leq 2$, this establishes $P(m)$. The result follows by induction.

15. *Proof by induction.* Let $P(n)$ be the statement, "One can make n-cents in postage using a combination of 3-cent and 8-cent stamps." We can check the first three statements as follows:

- $P(14)$ is true since $3 + 3 + 8 = 14$.
- $P(15)$ is true since $3 + 3 + 3 + 3 + 3 = 15$.
- $P(16)$ is true since $8 + 8 = 16$.

Now let $m \geq 17$ be given such that $P(m)$ is the first statement not yet checked. In particular, $P(m-3)$ has been checked, so we know that it is possible to make $m - 3$ cents in postage using just these types of stamps. Adding a 3-cent stamp to this postage consitutes m cents in postage, so $P(m)$ is true.

17. *Proof.* Let $P(n)$ be the statement,

$$\sum_{i=2}^{2^n} \frac{1}{i} \geq \frac{n}{2}$$

For example, $P(1)$ says, "$\frac{1}{2} \geq \frac{1}{2}$" which is true. Now let $m \geq 2$ be given such that $P(1), P(2), \ldots, P(m-1)$ have already been checked to be true. In particular, $P(m-1)$ says,

$$\sum_{i=2}^{2^{m-1}} \frac{1}{i} \geq \frac{m-1}{2}$$

We now consider the next statement $P(m)$:

$$\sum_{i=2}^{2^m} \frac{1}{i} = \left(\frac{1}{2} + \frac{1}{3} + \cdots + \frac{1}{2^{m-1}} \right) + \left(\frac{1}{2^{m-1}+1} + \frac{1}{2^{m-1}+2} + \cdots + \frac{1}{2^m} \right)$$
$$\geq \left(\frac{m-1}{2} \right) + \frac{1}{2} = \frac{m}{2}$$

The key step here is the fact that

$$\frac{1}{2^{m-1}+1} + \frac{1}{2^{m-1}+2} + \cdots + \frac{1}{2^m} \geq \underbrace{\frac{1}{2^m} + \frac{1}{2^m} + \cdots + \frac{1}{2^m}}_{2^{m-1} \text{ terms}} = \frac{1}{2}$$

This establishes that $P(m)$ is true, completing the induction.

19. In each case, the first error is given.

   (a) When $m = 1$, if the Reader picks the set $S = \varnothing$, then it is impossible to follow the instruction, "choose an element $a \in S$."

Section 2.5 exercises

1. *Proof.* Suppose a counterexample to this statement *does* exist. Let's agree to call it m. Since m is a counterexample, it must make the hypothesis of the original statement true while making the conclusion false. That is, $m = \underline{3K+1}$ for some integer K and $m = \underline{9L+5}$ for some integer L. Combining these equations gives us

$$3K+1 = 9L+5$$

from which it follows that $K - 3L = \frac{4}{3}$. Since $\underline{K - 3L}$ is an integer, this is nonsense. Therefore, no counterexample exists.

3. (a) *Proof.* Suppose a counterexample n to this statement exists. Since n makes the hypothesis true, $n^2 = 2K$ for some integer K. Since n makes the conclusion false, $n = 2L + 1$ for some integer L. Combining these equations gives us $2K = (2L+1)^2$. Using algebra, we find that $2K - 4L^2 - 4L = 1$, from which it follows that $K - 2L^2 - 2L = \frac{1}{2}$. Since $K - 2L^2 - 2L$ is an integer and $\frac{1}{2}$ is not, this contradiction tells us no counterexample exists.

   (b) *Proof.* Suppose a counterexample n to this statement exists. Since n makes the hypothesis true, $n^2 = 2K + 1$ for some integer K. Since n makes the conclusion false, $n = 2L$ for some integer L. Combining these equations gives us $2K+1 = (2L)^2$. Using algebra, we find that $2K - 4L^2 = -1$, from which it follows that $K - 2L^2 = -\frac{1}{2}$. Since $K - 2L^2$ is an integer and $-\frac{1}{2}$ is not, this contradiction tells us no counterexample exists.

5. *Proof by contradiction.* Suppose as a counterexample there are odd perfect squares a and b whose sum is the perfect square c. From Exericse 3(b), we know that a and b are the squares of odd integers. That is, $a = (2K + 1)^2$ and $b = (2L + 1)^2$ for some integers K and L. In this case,

$$c = a + b = 4K^2 + 4K + 1 + 4L^2 + 4L + 1$$
$$= 4(K^2 + K + L^2 + L) + 2$$

Hence, c is an even perfect square of the form $4M + 2$, where M is the integer $K^2 + K + L^2 + L$. This is impossible (by the previous exercise), so no such counterexample exists.

7. *Proof by contradiction* Suppose there is an integer n which is of the form $5K + 3$ and of the form $5L + 1$. for some integers K and L. This means that

$$5K + 3 = 5L + 1, \text{ or}$$
$$\frac{2}{5} = L - K$$

Since $L - K$ is an integer, this contradiction shows that there is no such integer n.

9. *Proof by contradiction.* Suppose to the contrary that there *are* positive integers a and b, with no common divisors, satisfying $a^2 = 2b^2$. Since a and b have no common divisor greater than 1, they cannot both be even. Thus either: (1) a is even and b is odd; or (2) b is even and a is odd; or (3) both are odd. We can proceed by cases.

- **Case 1.** If a is even and b is odd, we write $a = 2m$ and $b = 2n + 1$ for integers m and n. Substituting into $a^2 = 2b^2$ and simplifying, we find that $m^2 - 2n^2 - 2n = \frac{1}{2}$, which is a contradiction to the closure of the integers under the operations on the left-hand side.

- **Case 2.** If a is odd and b is even, we write $a = 2m + 1$ and $b = 2n$ for integers m and n. Substituting into $a^2 = 2b^2$ and simplifying, we find that $2n^2 - m^2 - m = \frac{1}{4}$, which is a contradiction to the closure of the integers under the operations on the left-hand side.

- **Case 3.** If a and b are both odd, we write $a = 2m + 1$ and $b = 2n + 1$ for integers m and n. Substituting into $a^2 = 2b^2$ and simplifying, we find that $m^2 + m - 2n^2 - 2n = \frac{1}{4}$, which is a contradiction to the closure of the integers under the operations on the left-hand side.

In every case, a contradiction arises. Hence our original assumption that there are positive integers a and b, with no common divisors, satisfying $a^2 = 2b^2$, must be incorrect.

10. (a) Suppose a positive number x divided by a positive number y results in a negative number z. Since $x \div y = z$ implies that $x = y \cdot z$, this means that the positive number x is the product of a positive number and a negative number, contradicting "rule" (ii). Hence, z must be positive.

11. Suppose, to the contrary, that there is a rational number a such that $a + \sqrt{2}$ is rational. Then $(a + \sqrt{2}) - a$ is rational. But this means $\sqrt{2}$ is rational, a contradiction of Theorem 4. Hence, there is no such rational number a.

13. The contrapositive is, "If $a + b$ is rational, then a is irrational or b is rational."

Proof. Let a and b be given such that $a + b$ is rational. Either a is irrational or a is rational.

Case 1. If a is irrational, then the conclusion, "a is irrational or b is rational" is certainly true.

Case 2. If a is rational, then by Exercise 9 in Section 2.2, $(a + b) - a = b$ is rational, and hence the conclusion, "a is irrational or b is rational" is also true.

Since in either case the same conclusion holds, it must be true that a is irrational or b is rational.

14. (a) *Proof.* Since $6\left(\frac{5}{3}\right)^2 + 11\left(\frac{5}{3}\right) = \frac{50}{3} + \frac{55}{3} = 35$, it follows that there exists a positive rational number r such that $6r^2 + 11r = 35$.

16. The contrapositive of Proposition 3 in this exercise is, "If r is rational, then $r^2 \neq 2$."

Proof. Let the rational number r be given. By Proposition 1, $r = \frac{a}{b}$ for integer a and b having no common divisor greater than 1. This means that a and b do not have 2 as a common divisor. By the contrapositive of Proposition 2, this means that $(\frac{a}{b})^2 \neq 2$. That is, $r^2 \neq 2$, as desired.

18. (a) For every integer n, n is not the largest integer.

(b) Suppose we are given an integer n. To show that n is not the largest integer, all we have to do is find a larger integer – and $n + 1$ certainly fits that description.

19. The contrapositive statement is, "If n is even, then $5n + 4$ is even".

Proof. Let an even integer n be given. Since n is even, there is an integer k such that $n = 2k$. In this case,

$$5n + 4 = 5(2k) + 4$$
$$= 2(5k + 2)$$

Since $5k + 2$ is an integer, this means that $5n + 4$ is even.

23. The contrapositive statement is, "If each of 7 boxes contains less than 17 ounces, then the average is not 17 ounces."

Proof. For each i, let c_i be the number of ounces in cereal box i. Then the hypothesis can be stated, "$c_i < 17$ for each i." From this it follows that the average is

$$\frac{c_1 + c_2 + c_3 + c_4 + c_5 + c_6 + c_7}{7} < \frac{17 + 17 + 17 + 17 + 17 + 17 + 17}{7} = 17$$

That is, the average of the c_i's is less than 17, hence it is not equal to 17.

25. The contrapositive is, "If every number x in a collection satisfies $x < m$, then the collection does not have an average of m."

Proof. Let the collection $x_1, x_2, \ldots, x_n$ satisfying the hypothesis be given. Since $x_1 < m, x_2 < m, \ldots$, and x_n, m, it follows that

$$x_1 + x_2 + \cdots + x_n < \underbrace{m + m + \ldots + m}_{n \text{ times}}$$
$$= m \cdot n$$

So the average of the collection is

$$\frac{x_1 + x_2 + \cdots + x_n}{n} < \frac{m \cdot n}{n} = m.$$

Hence the average of the collection is less than (and consequently, not equal to) m, as desired.

26. Pigeonhole Principle (Basic Version): If $n + 1$ objects are distributed among n boxes, then some box must contain more than one object.

Proof. Let $n + 1$ objects be given, and label the n boxes $1, 2, 3, \ldots, n$. After the objects have been distributed among the boxes, define $x_1, x_2, \ldots, x_n$ by the following rule:

$$x_i = \text{ the number of objects in box } i$$

Since each object can go into only one box, we know that

$$x_1 + x_2 + x_3 + \cdots + x_n = n + 1$$

which means the average value of the x's is

$$\frac{x_1 + x_2 + x_3 + \cdots + x_n}{n} = \frac{n + 1}{n}$$
$$= 1 + \frac{1}{n}$$

According to the Average Version, there is an x_i that is at least $1 + \frac{1}{n}$. Since x_i is an integer value, this means that $x_i > 1$, which can be interpreted as meaning, "more than 1 object is in the box labeled i."

Pigeonhole Principle (General Version): If $m \cdot n + 1$ objects are distributed among n different boxes, then there must be some box containing at least $m + 1$ objects.

Proof. Let $m \cdot n + 1$ objects be given, and label the n boxes $1, 2, 3, \ldots, n$. After the objects have been distributed among the boxes, define $x_1, x_2, \ldots, x_n$ by the following rule:

$$x_i = \text{the number of objects in box } i$$

Since each object can go into only one box, we know that

$$x_1 + x_2 + x_3 + \cdots + x_n = m \cdot n + 1$$

which means the average value of the x's is

$$\frac{x_1 + x_2 + x_3 + \cdots + x_n}{n} = \frac{m \cdot n + 1}{n}$$
$$= m + \frac{1}{n}$$

According to Exercise 25, there is an x_i that is at least $m + \frac{1}{n}$. Since x_i is an integer value, this means that $x_i \geq m + 1$, which can be interpreted as meaning, "at least $m + 1$ objects are in the box labeled i."

27. Statement of contrapositive. If integers x, y, and z satisfy $x < 4$, $y < 4$, and $z < 5$, then $x + y + z < 11$.

Proof. Since $x < 4$, $y < 4$, and $z < 5$, and since x, y, and z are integers, we know that $x \leq 3$, $y \leq 3$, and $z \leq 4$. So $x + y + z \leq 3 + 3 + 4$, that is $x + y + z \leq 10$. Since $10 < 11$, we conclude that $x + y + z < 11$.

30. We can write the basic version as: Objects are distributed among n boxes. If $n + 1$ objects are distributed, then some box must contain more than one object.

Contrapositive: If no box contains more than one object, then the number of object distributed is not $n + 1$.

Proof. After the objects have been distributed among the boxes, define $x_1, x_2, \ldots, x_n$ by the following rule:

$$x_i = \text{the number of objects in box } i$$

Since no box contains more than one object, we know that each x_i satisfies $x_i \leq 1$. Thus the total number of objects distributed is given by

$$x_1 + x_2 + x_3 + \cdots + x_n \leq n$$

This shows that the total cannot be $n + 1$.

31. This statement is similar to Exercise 25: "For any list of numbers, if one of the numbers is greater than the average, then at least one of the numbers must be less than the average." The proof of the statement is related to the Average Version of the Pigeonhole Principle. However, its truth is intuitively obvious to almost everyone, and that is why Garrison Keillor's line is funny to almost everyone. (Some of today's bureaucrats don't get the joke.)

32. (a) *Proof.* Define Boxes 0, 1, 2, 3, 4 and 5 by the following rule: "Place integer x into Box r of x is of the form $6q + r$." The Division Theorem tells us that every integer can be placed in this way. Let seven integers be given. By the Pigeonhole Principle, some box must contain (at least) two integers. Let's call the two integers x and y, and say they are in Box d. The rule defining the boxes tells us that $x = 6 \cdot K + d$ and $y = 6 \cdot L + d$ for some integers K and L. In this case,

$$x - y = (6K + d) - (6L + d)$$
$$= 6 \cdot (K - L).$$

Since $K - L$ is an integer, this means that $x - y$, the difference between x and y, is divisible by 6.

33. *Proof.* Let five integers be given. Think of two boxes, one labeled "divisible by 3" and one labeled "not divisible by 3." By the distribution version of the Pigeonhole Principle (with $n = \underline{3}$ and $m = \underline{2}$), we conclude that there are at least three of the numbers in one box. Let's refer to these three numbers as a, b and c, and consider two cases based on which box they are in.

- **Case 1.** If a, b, c are in the box labeled "divisible by 3," then $a^2 + b^2 + c^2$ is divisible by 3 because $a = 3K$, $b = 3L$ and $c = 3J$ (where K, L, and J are integers) imply that

$$a^2 + b^2 + c^2 = (3K)^2 + (3L)^2 + (3J)^2$$
$$= 3(3K^2 + 3L^2 + 3J^2)$$

which is certainly divisible by 3.

- **Case 2.** If a, b, c are in the box labeled "not divisible by 3," then by Practice Problem 4 from Section 2.2, a^2 can be written in the form $\underline{3K + 1}$, b^2 can be written in the form $\underline{3L + 1}$, and c^2 can be written in the form $\underline{3J + 1}$. Hence, $a^2 + b^2 + c^2$ is divisible by 3 because in this case,

$$a^2 + b^2 + c^2 = (3K + 1) + (3L + 1) + (3J + 1)$$
$$= 3(K + L + J + 1)$$

In either case, $a^2 + b^2 + c^2$ is divisible by 3, completing the proof.

35. (a) Considering the four triangular regions shown to be "boxes", any five points will require two to share one box by the Pigeonhole Principle.

The greatest distance between two points in one of these small triangular regions is the distance between vertices, which is $\frac{1}{2}$. Hence, there must be two points within $\frac{1}{2}$ of each other.

36. (c) If $K\pi$ is in Box 1, then $K\pi$ is within $\frac{1}{n}$ of an integer.

(d) If no multiple of π is in Box 1, then all n given multiples are distributed among $n - 1$ boxes, so the Pigeonhole Principle guarantees that some box will contain (at least) two multiples.

(e) If $K\pi$ and $L\pi$ are both in the same box (where $K > L$, say), then $K\pi - L\pi$ has a fractional part between $\frac{n-1}{n}$ and 1 or between 0 and $\frac{1}{n}$.

(f) In either case described in (e), $K\pi - L\pi$ is within $\frac{1}{n}$ of an integer. Since $(K - L)\pi$ is one of the multiples originally described, this is the desired conclusion.

Section 2.6 Exercises

1. (a) 100011
   (c) 1111011

2. (a) 120
   (c) 443

3. (a) 43
   (c) 173

4. (a) 29
   (c) 10

5. For convenience in reading them, we show the binary numbers in groups of 4 bits.

   (a) $(1101\ 1010\ 1101)_2$
   (b) $(1\ 1111\ 0000\ 1011)_2$

(d) $(19)_{16}$

(e) $(B0DE)_{16}$

6. For convenience in reading them, we show the binary numbers in groups of 3 bits.

(a) $(1\ 111)_2$

(b) $(10\ 000\ 000\ 101)_2$

(d) $(31)_8$

(e) $(130336)_8$

7. (a) $(1111)_2 = (F)_{16}$

(b) $(1010100101)_2 = (2A5)_{16}$

(d) $(110110101101)_2 = (6655)_8$

(e) $(100010001000100010001)_2 = (4210421)_8$

8. Since this is an "if and only if" proposition, there are two separate proofs to be completed.

- **Claim 1.** If x is divisible by 3, then the sum of the decimal digits of x is divisible by 3.

  Proof. Let a natural number x which is divisible by 3 be given, and let s be the sum of the decimal digits of x. Since x is divisible by 3, we know that $x = 3K$ for some integers. By Proposition 3, we know that $x - s$ is divisible by 9, so $x - s = 9L$ for some integer L. From this it follows that $3K - 9L = s$, so $s = 3(K - 3L)$. Hence, s is divisible by 3.

- **Claim 2.** If the sum of the decimal digits of x is divisible by 3, then x is divisible by 3.

  Proof. Let a natural number x whose digits sum to a number s that is divisible by 3 be given. That is, $s = 3K$ for some integer K. By Proposition 3, we know that $x - s = 9L$ for some integer L, and so $x = s + 9L = 3K + 9L = 3(K + 3L)$. Hence, x is divisible by 3.

9. 0,1 or 4

11. (a) and (c)

12. Let the integer n be represented as $d_4d_3d_2d_1d_0$ in base b. This means that

$$
\begin{aligned}
n &= d_4 \cdot b^4 + d_3 \cdot b^3 + d_2 \cdot b^2 + d_1 \cdot b + d_0 \text{ and so} \\
n \cdot b &= d_4 \cdot b^5 + d_3 \cdot b^4 + d_2 \cdot b^3 + d_1 \cdot b^2 + d_0 \cdot b + 0
\end{aligned}
$$

Hence, $n \cdot b$ is represented as $d_4d_3d_2d_1d_00$ in base b.

13. Nine is divisible by 3, but its binary representation $(1001)_{two}$ has digits that sum to 2.

14. Seven is not divisible by 3 but its binary representation $(111)_{two}$ has digits that sum to 3.

17. **Proposition 1.** For all $n \geq 0, 8^n - 1$ is divisible by 7.

Proof by induction on n. Let $P(n)$ be the statement, "$8^n - 1$ is divisible by 7". The first statement is $P(0)$, which says, "$8^0 - 1$ is divisible by 7". Since $0 = 7 \cdot 0$, this is clearly true. Now let $m \geq 1$ be given such that the statements $P(0), P(1), \ldots, P(m-1)$ have been checked, and we are ready to consider statement $P(m)$. Since statement $P(m-1)$ has been checked, we know that $8^{m-1} - 1 = 7 \cdot K$ for some integer K. Then

$$
\begin{aligned}
8^m - 1 &= 8(8^{m-1} - 1) + 7 \\
&= 8 \cdot (7K) + 7 \\
&= 7 \cdot (8K + 1)
\end{aligned}
$$

Hence, $8^m - 1$ is divisible by 7, so statement $P(m)$ is also true. $\qquad\square$

Proposition 2. If s is the sum of the digits in the octal representation of x, then $x - s$ is divisible by 7.

Proof. Let a natural number x be given, and write the octal representation as $(d_K d_{K-1} \ldots d_1 d_0)_{\text{octal}}$. This means that

$$
\begin{aligned}
x &= d_K \cdot 8^K + d_{K-1} \cdot 8^{K-1} + \cdots + d_1 \cdot 8^1 + d_0 \cdot 8^0 \\
&= \sum_{i=0}^{K} d_i \cdot 8^i
\end{aligned}
$$

Now let s be the sum of the octal digits of x. That is,

$$
\begin{aligned}
s &= d_K + d_{K-1} + \ldots + d_1 + d_0 \\
&= \sum_{i=0}^{K} d_i
\end{aligned}
$$

Hence,

$$
\begin{aligned}
x - s &= \sum_{i=0}^{K} d_i \cdot 8^i - \sum_{i=0}^{K} d_i \\
&= \sum_{i=0}^{K} d_i \cdot (8^i - 1)
\end{aligned}
$$

Since each $8^i - 1$ is divisible by 7 (by Proposition 1), it follows that $x - s$ is divisible by 7.

To prove the "if and only if" statement of 17, we must prove two separate things.

- **Claim 1**. If a natural number x is divisible by 7, then the sum of its octal digits is divisible by 7.

  Proof. Let a natural number x divisible by 7 be given, and let s be the sum of its octal digits. Since x is divisible by 7, then $x = 7 \cdot K$ for some integer K. By Proposition 2 above, $x - s = 7 \cdot L$ for some integer L. Hence, $7 \cdot K - s = 7L$, or $s = 7(K - L)$. Since, $K - L$ is an integer, this shows that s is divisible by 7.

- **Claim 2**. If a natural number x has the sum of its octal digits divisible by 7, then x is divisible by 7.

  Proof. Let a natural number x be given such that the sum of its octal digits (let's call this sum s) is divisible by 7. That is, $s = 7K$ for some integer K. By Proposition 2 above, $x - s = 7L$ for some integer L. Hence, $x - 7K = 7L$, or $x = 7(K + L)$. Since $K + L$ is an integer, this shows that x is divisible by 7.

19. (b) #FFFF00 is yellow. (c) Any value from $(000000)_{\text{hex}}$ to $(FFFFFF)_{\text{hex}}$ will work. $(000000)_{\text{hex}}$ is the hexadecimal representation of 0, and

$$
\begin{aligned}
(FFFFFF)_{\text{hex}} &= 15 \cdot 16^5 \cdot 15 \cdot 16^4 + 15 \cdot 16^3 + 15 \cdot 16^2 + 15 \cdot 16 + 15 \\
&= 16777215
\end{aligned}
$$

so there are 16777216 different background colors to choose from.

20. We use T for ten and E for eleven.

    (a) 4321

    (c) $3T1$

Section 2.7 Exercises

1. (a) $a = 2, b = 2, n = 4$

(c) $a = 2$, $n = 8$

(e) A trivial example is $n = 2$ since $1^2 \equiv_2 5$ and 5 does not divide $2 \cdot 1$. A more substantial example is $n = 19$ noting that $9^2 \equiv_{19} 5$, but 5 does not divide $19 \cdot 18$.

2. (a) **Proposition.** For any integer $n \neq 0$ and for all integers a, b, c and d, if $a \equiv_n b$ and $c \equiv_n d$, then $a + c \equiv_n b + d$

Proof. Let a, b, c, d and n be given, and assume that $a \equiv_n b$ and $c \equiv_n d$. By Theorem 1, both $a - b$ and $c - d$ are divisible by n. *This means $a - b = Kn$ and $c - d = Ln$ for some integers K and L. Adding these equations, we get*

$$a - b + c - d = Kn + Ln$$

which can be rearranged as $(a + c) - (b + d) = (K + L) \cdot n$. Since $K + L$ is an integer, this means that $(a + c) - (b + d)$ is divisible by n. Since $(a + c) - (b + d)$ is divisible by n, we conclude (by Theorem 1 again) that $a + c \equiv_n b + d$.

3. The mod 7 tables are given below.

+	0	1	2	3	4	5	6
0	0	1	2	3	4	5	6
1	1	2	3	4	5	6	0
2	2	3	4	5	6	0	1
3	3	4	5	6	0	1	2
4	4	5	6	0	1	2	3
5	5	6	0	1	2	3	4
6	6	0	1	2	3	4	5

$\cdot$	0	1	2	3	4	5	6
0	0	0	0	0	0	0	0
1	0	1	2	3	4	5	6
2	0	2	4	6	1	3	5
3	0	3	6	2	5	1	4
4	0	4	1	5	2	6	3
5	0	5	3	1	6	4	2
6	0	6	5	4	3	2	1

5. In mod 11 arithmetic, we have

$$1^{-1} = 1, 2^{-1} = 6, 3^{-1} = 4, 5^{-1} = 9, 7^{-1} = 8, 10^{-1} = 10$$

8. Proposition 3 tells us that $x - s_x$ is divisible by 9. This means $x - s_x = 9k$ for some integer k, which means $x \equiv_9 s_x$.

10. (a) Every $x \equiv_{15} 2$ will work.

(c) Every $x \equiv_{11} 4$ or $x \equiv_{11} 8$ will work.

(e) Every $x \equiv_{17} 4$ or $x \equiv_{17} 13$ will work.

11. p can be any element of $\{2, 5, 13, 17, 29, 37, 41, 53, 61, 73, 89, 97\}$. Except for 2, these are the primes of the form $4K + 1$, where K is an integer.

13. This is equivalent to the claim, "$n^5 \equiv_{10} n$ for all $n \in \mathbb{N}$." *Proof.* Let $n \in \mathbb{N}$ be given. By the Division Theorem, $n \equiv_{10} d$ for some $d \in \{0, 1, 2, 3, 4, 5, 6, 7, 8, 9\}$ (d is the ones' digit of n), so we can argue in ten cases:

Case 0. If $n \equiv_{10} 0$, then $n^5 \equiv_{10} 0$.

Case 1. If $n \equiv_{10} 1$, then $n^5 = 1 \equiv_{10} 1$.

Case 2. If $n \equiv_{10} 2$, then $n^5 = 32 \equiv_{10} 2$.

Case 3. If $n \equiv_{10} 3$, then $n^5 = 243 \equiv_{10} 3$.

Case 4. If $n \equiv_{10} 4$, then $n^5 = 1024 \equiv_{10} 4$.

Case 5. If $n \equiv_{10} 5$, then $n^5 = 3125 \equiv_{10} 5$.

Case 6. If $n \equiv_{10} 6$, then $n^5 = 7776 \equiv_{10} 6$.

Case 7. If $n \equiv_{10} 7$, then $n^5 = 16807 \equiv_{10} 7$.

Case 8. If $n \equiv_{10} 8$, then $n^5 = 32768 \equiv_{10} 8$.

Case 9. If $n \equiv_{10} 9$, then $n^5 = 59049 \equiv_{10} 9$.

In each case, $n^5 \equiv_{10} n$.

14. (a) Since $k = (p-1)(q-1) = 60$ and $e = 43$, we can determine that

$$7 \cdot 43 + (-5) \cdot 60 = 1$$

and so $d = 7$ will work as a decryption key.

(b) We apply this decryption value (along with public key $n = 77$) to $[41, 26, 26, 69, 1, 69, 41, 64, 26, 61]$ as follows:

- $41^7 \bmod 77 = 13$ (M)
- $26^7 \bmod 77 = 5$ (E)
- $69^7 \bmod 77 = 20$ (T)
- $1^7 \bmod 77 = 1$ (A)
- $64^7 \bmod 77 = 15$ (O)
- $61^7 \bmod 77 = 19$ (S)

So the message is $[M, E, E, T, A, T, M, O, E, S]$ — "Meet at Moe's."

(c) (1) Making $n = 77$ and $e = 43$ publically available, makes it easy to determind $p = 7$ and $q = 11$ and hence that $k = 60$. Form this, it is pretty easy to find the decryption key. Hence, using a small value of n is not very secure. (2) Encrypting one character at a time means there will be a lot of repetition of numbers in your message. A long message can be easily broken using knowledge about common letters and letter patterns. Hence, encrypting one letter at a time can be broken without using RSA methods at all.

17. (a) A computer search yields 341, 561, and 645 as the only three possibilities.

(c) A computer search yields 124, 217, 561, and 781 as the only four possibilities.

18. (a) *Proof.* Let $n \equiv_6 5$ be given. This means that $n = 6k + 5$ for some $k \in \mathbb{Z}$. But then $n = 2(3k + 2) + 1$ and $n = 3(2k + 1) + 2$. That is, $n \equiv_2 1$ and $n \equiv_3 2$.

19. (a) $x = 43$

(c) No solution.

(e) No solution.

Section 3.1 exercises

1. (a) $\{2\}$

(b) $\{1, 2, 4, 8\}$

(c) $\{3, 4, 7, 8, 9\}$

(d) $\{8\}$

(e) $\{1, 2, 5, 6, 10\} \cap \{3, 4, 5, 6, 7, 9, 10\} = \{5, 6, 10\}$

3. (a) $\{x \in \mathbb{Z} : x = 2y \text{ for some } y \in \mathbb{Z}\}$

(b) $\{x \in \mathbb{N} : x = 2^y \text{ for some } y \in \mathbb{N}\}$

(c) $\{x \in \mathbb{N} : x = 2y + 1 \text{ for some } y \in \mathbb{N}\}$

(d) $\{x \in \mathbb{N} : x = (2y + 1)^2 \text{ for some } y \in \mathbb{N}\}$

5. (a) $\{7, 13, 19, 25, 31, 37, 43, \ldots\}$

(b) $\{\ldots, -17, -11, -5, 1, 4, 7, 10, 13, 16, \ldots\}$

(c) $\{3, 5, 9, 11, 15, 17, 21, 23, 27, \ldots\}$

(d) $\{4, 10, 16, 22, 28, 34, 40, 46, 52, \ldots\}$

7. (a) $\{2x : x \in \mathbb{Z}\}$

(b) $\{x^3 : x \in \mathbb{Z}\}$

(c) $\{10x + 7 : x \in \mathbb{N}\}$

(d) $\{\frac{a}{b} \in \mathbb{Q} : -b < a < b\}$

8. (a) $\not\subseteq$ because $0 \in \mathbb{N}$ and $0 \notin \mathbb{Z}^+$

(c) $\not\subseteq$ because $-\frac{1}{2} \in \mathbb{Q}$ and $-\frac{1}{2} \notin \mathbb{R}^+$

(e) $\subseteq$

9. (a) B since $3 - 4 = -1 \in B$

10. (a) $\{2n + 1 : n \in \mathbb{Z}\}$

(c) $\{2y^2 : y \in \mathbb{N}\}$

11. (a) $\{\frac{2}{3}, \frac{4}{5}, \frac{6}{7}, \frac{8}{9}, \frac{10}{11}, \ldots\}$ with universe $\mathbb{Q}$

(c) $\{1, 2, \frac{1}{2}, 4, \frac{1}{4}, 9, \frac{1}{9}, 27, \frac{1}{27}, \ldots\}$ with universe $\mathbb{Q}$

12. (a) These are all positive odd numbers, $\{2x + 1 : x \in \mathbb{N}\}$.

(c) These are each one more than a multiple of 8, $\{8x + 1 : x \in \mathbb{N}\}$

13. (a) True.

(b) False. $A = \{1, 2\}, B = \{a, b\}$

(c) True.

(d) False. $A = \{1, 2\}, B = \{a, b, c\}$

15. (a) $A \cap (B \cup C) = (A \cap B) \cup (A \cap C) = \{1, 3, 5\}$

(c) $(A \cup B)' = A' \cap B' = \{6, 7, 8, 9, 10\}$

(e) $A \cap (A \cup B) = A = \{1, 3, 5\}$

16. Only the final Venn diagrams illustrating each property is shown:

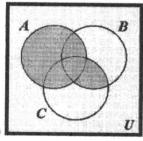 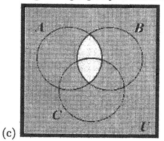

(a) (c)

17. The Venn diagram for each side of the equation is given. For each that do not match, an example is given to illustrate the difference.

(a)

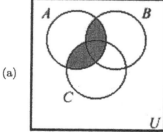 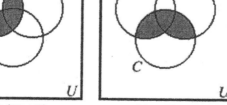

$A \cap (B \cup C)$ $(A \cup B) \cap C$

Counterexample: $A = \{1, 2, 3\}, B = \{1, 3, 4\}, C = \{1, 2, 4\}$

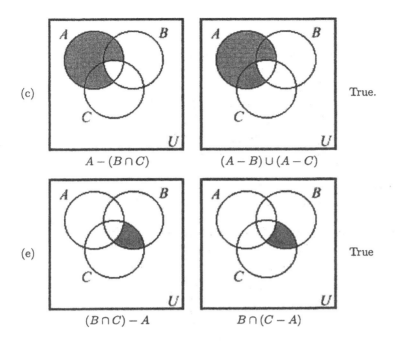

(c)	True.
$A - (B \cap C)$ $(A - B) \cup (A - C)$	
(e)	True
$(B \cap C) - A$ $B \cap (C - A)$	

19. (a) Any partial sum of the harmonic series, in lowest terms, has an even denominator.

    (b) Any partial sum of the harmonic series is between 1 and 2.

    (c) There is at least one rational number between 1 and 2 with an even denominator (in lowest terms).

    (d) $C \subseteq A$

20. (a) $\{-5, -4, -3, -2, -1, 0, 1, 2, 3\}$

    (b) $[3.1, 4]$

    (c) $\emptyset$

22. (a) $[0, 12]$

    (c) $[2, 27]$

23. (a) $n(A) = 8$, $n(B) = 9$, $n(C) = 12$

    (b)   i. $n(A \cap B) = n(\{2, 9, 12, 13\}) = 4$, $n(A \cup B) = n(A) + n(B) - n(A \cap B) = 8 + 9 - 4 = 13$

       ii. $n(A \cap C) = n(\{7, 9, 12, 13, 16\}) = 5$, $n(A \cup C) = n(A) + n(C) - n(A \cap C) = 8 + 12 - 5 = 15$

      iii. $n(B \cap C) = n(\{4, 8, 9, 11, 12, 13, 14\}) = 7$, $n(B \cup C) = n(B) + n(C) - n(B \cap C) = 9 + 12 - 7 = 14$

      iv. $n(A \cap B \cap C) = n(\{9, 12, 13\}) = 3$

       v. $n(A \cup B \cup C) = n(\{1, 2, 3, 4, 5, 6, 7, 8, 9, 10, 11, 12, 13, 14, 15, 16\}) = 16$

26. (a) Let $A = \{n \in T : n \text{ is a multiple of } 2\}$ and $B = \{n \in T : n \text{ is a multiple of } 3\}$. Then

$$n(A \cup B) = n(A) + n(B) - n(A \cap B) = 501 + 334 - 167 = 668$$

27. The Venn diagram is followed by the answers to the questions.

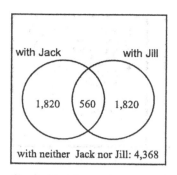

(a) $1820 + 1820 + 560 = 4200$

(c) 560

29. As the diagram shows, there are three students that have completed all three stations.

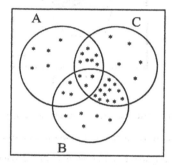

32. (a) True. Since there is no overlap, the "$-B$" does not remove anything from A.

(c) False. $A = \{1\}, B = \{2\}$

Section 3.2 exercises

1. (a) $\{(2,1),(2,2),(2,8),(4,1),(4,2),(4,8)\}$

   (c) $\{\{\}, \{1\}, \{2\}, \{8\}, \{1,2\}, \{1,8\}, \{2,8\}, \{1,2,8\}\}$

3. (a) 4; since $\{S \in P(\{1,2,3\}) : n(S) \geq 2\} = \{\{1,2\}, \{1,3\}, \{2,3\}, \{1,2,3\}\}$

   (b) 6; since $\{S \in P(\{1,2,3\}) : S \cap \{1,2\} \neq \emptyset\} = \{\{1\}, \{2\}, \{1,2\}, \{1,3\}, \{2,3\}, \{1,2,3\}\}$

   (c) 4; since $\{S \in P(\{1,2,3,4\}) : S \cap \{1,4\} = \emptyset\} = \{\emptyset, \{2\}, \{3\}, \{2,3\}\}$

4. (a) $(\{1\}, \{1,2\}), (\{2\}, \{2,3\}), (\emptyset, \{1\}), (\{1,2\}, \{1,2,3\}), (\{2,3\}, \{1,2,3\})$

5. $A = \{4k : k \in \mathbb{Z}\} = \{0, \pm 4, \pm 8, \pm 12, \pm 16, \pm 20, \ldots\}$

   $B = \{4k + 1 : k \in \mathbb{Z}\} = \{\ldots, -19, -15, -11, -7, -3, 1, 5, 9, 13, 17, 21, \ldots\}$

   $C = \{4k + 2 : k \in \mathbb{Z}\} = \{\ldots, -14, -10, -6, -2, 2, 6, 10, 14, 18, 22, \ldots\}$

   $\{A, B, C\}$ is not a partition of $\mathbb{Z}$ since $A \cup B \cup C \neq \mathbb{Z}$; for example, $3 \notin A \cup B \cup C$.

   Also $7 \notin A \cup B \cup C$. In fact $4k + 3 \notin A \cup B \cup C, \forall k \in \mathbb{Z}$. Note however: $\{A, B, C\}$ are pairwise disjoint.

7. $5 \notin A \cup B \cup C \cup D$ and so $A \cup B \cup C \cup D \neq \mathbb{Z}$

9. We know the three sets are non-empty since $0 \in A$, $1 \in B$ and $3 \in C$. The Division theorem tells us that every integer can be written in one of the forms, $4k$, $4k + 1$, $4k + 2$ or $4k + 3$. This means that any given integer is either in A (if it's of the form $4k$ or $4k + 2$) or in B (if it's in the form $4k + 1$) or in C (if it's of the form $4k + 3$). No pair of these sets overlap, as can be shown with a simple proof by contradiction. For

example, to show that $A \cap B = \emptyset$, we simpy assume that there *is* an $x \in A \cap B$, and argue that $x = 2K$ and $x = 4L + 1$ for some integers K and L. From this, it will follow that $K - 2L = \frac{1}{2}$, a contradiction to the closure of the integers.

10. (a) Let $A = \{1\}, B = \{2\}, C = \{3\}$. Then $A \times (B \cup C) = \{(1,2), (1,3)\}$, whereas $(A \times B) \cap (A \times C) = \emptyset$

(c) Let $A = \{1,2\}, B = \{2,3\}$. Then $P(A) \cup P(B) = \{\emptyset, \{1\}, \{2\}, \{1,2\}, \{3\}, \{2,3\}\}$, whereas $P(A \cup B) = \{\emptyset, \{1\}, \{2\}, \{3\}, \{1,2\}, \{1,3\}, \{2,3\}, \{1,2,3\}\}$

13. (a) $n(P(A)) = 2^{n(A)} = 2^3 = 8$

(b) $n(P(B)) = 2^{n(B)} = 2^4 = 16$

(c) $n(A \times B) = n(A) \cdot n(B) = 3 \cdot 4 = 12$

(d) $n(P(A \times B)) = 2^{n(A \times B)} = 2^{n(A) \cdot n(B)} = 2^{3 \cdot 4} = 2^{12} = 4096$

(e) $n(P(A) \times P(B)) = 2^{n(A)} \cdot 2^{n(B)} = 2^3 \cdot 2^4 = 2^7 = 128$

15. There are ten partitions of the set $\{A, B, C, D, E, F\}$, having two parts, each containing three elements:

$\{\{A,B,C\}, \{D,E,F\}\}$ $\{\{A,C,E\}, \{B,D,F\}\}$

$\{\{A,B,D\}, \{C,E,F\}\}$ $\{\{A,C,F\}, \{B,D,E\}\}$

$\{\{A,B,E\}, \{C,D,F\}\}$ $\{\{A,D,E\}, \{B,C,F\}\}$

$\{\{A,B,F\}, \{C,D,E\}\}$ $\{\{A,D,F\}, \{B,C,E\}\}$

$\{\{A,C,D\}, \{B,E,F\}\}$ $\{\{A,E,F\}, \{B,C,D\}\}$

16. (a) Not a partition since not every element in S (like 1 and 2) is a subset of A.

(c) Not a partition since 6 is not an element of any part.

17. In each case, there is more than one correct answer.

(a) $\{\{1,2,3,4\}, \{5,6,7,8\}\}$.

(c) $\{\{2,4,6,8\}, \{1,3,5,7\}\}$

19. *Proof.* Let $S(n)$ be the statement, "If $n \geq 1$ is an integer, $A = \{1,2,3,4,5\}, B = \{1,2,3,\ldots,n\}$, then $n(A \times B) = 5n$." Statement $S(1)$ is true since

$$A \times B = \{1,2,3,4,5\} \times \{1\} = \{(1,1), (2,1), (3,1), (4,1), (5,1)\}$$

Now let $m \geq 2$ be given such that statements $S(1), S(2), \ldots, S(m-1)$ have all been verifies to be true. In particular, $S(m-1)$ states that $\{1,2,3,4,5\} \times \{1,2,\ldots,m-1\}$ is a set containing $5(m-1)$ elements/ordered pairs. Now consider the next statement, $S(m)$. The set/cartesian product $\{1,2,3,4,5\} \times \{1,2,\ldots,m-1,m\}$ clearly contains all of the elements/ordered pairs in the set $\{1,2,3,4,5\} \times \{1,2,\ldots,m-1\}$, as well as the five elements $(1,m), (2,m), (3,m), (4,m)$ and $(5,m)$, for a total of $5(m-1) + 5 = 5m$ elements. This establishes statement $S(m)$, completing the induction.

21. Let $A = \{1,2\}, B = \{b_1, b_2\}, C = \{x\}$. Then, $A \times B \times C = \{(1,b_1,x), (1,b_2,x), (2,b_1,x), (2,b_2,x)\}$, and $(A \times B) \times C = \{((1,b_1),x), ((1,b_2),x), ((2,b_1),x), ((2,b_2),x)\}$. The difference between these sets is that the first is a set of ordered 3-tuples, whereas the second set contains ordered pairs for each of which, the first coordinate is itself an ordered pair.

Section 3.3 exercises

1. (a) *Proof.* Dear READER, Remember that $\mathbb{N} = \{0,1,2,3,\ldots\}$ and that a natural number x is odd if and only if it can be written as $x = 2y + 1$, for some $y \in \mathbb{N}$. Now, you must agree that each of $1,3,5,7$ and 9 is a natural number. Also, note that $1 = 2 \cdot 0 + 1, 3 = 2 \cdot 1 + 1, 5 = 2 \cdot 2 + 1, 7 = 2 \cdot 3 + 1$ and $9 = 2 \cdot 4 + 1$, which establishes that each of $1,3,5,7$ and 9 is an odd number. Hence, each element of $\{1,3,5,7,9\}$ is an element of $\{k \in \mathbb{N} : k \text{ is odd}\}$; that is, $\{1,3,5,7,9\} \subseteq \{k \in \mathbb{N} : k \text{ is odd}\}$.

(c) This is false since 2 is a prime number that is not odd.

2. (a) *Proof.* Let $x \in \{4m : m \in \mathbb{Z}\}$, so that $x = 4m$, for some $m \in \mathbb{Z}$. We can write $x = 4m = 2(2m)$. Since $2m \in \mathbb{Z}$, $x \in \{2n : n \in \mathbb{Z}\}$, and so $\{4m : m \in \mathbb{Z} \subseteq \{2n : n \in \mathbb{Z}\}$.

(c) *Proof.* Let $x \in \mathbb{Z}$. We can write $x = \frac{x}{1}$. Since $x \in \mathbb{Z}$ and $1 \in \mathbb{Z}$ (obviously $1 \neq 0$), it follows that $x \in \mathbb{Q}$. Hence, $\mathbb{Z} \subseteq \mathbb{Q}$.

(e) *Proof.* Let Let $x \in \{2n + 1 : n \in \mathbb{Z}\} \cap \{5m + 4 : m \in \mathbb{Z}\}$, so that $x = 2n + 1$ and $x = 5m + 4$, for some $n, m \in \mathbb{Z}$. Now, $2n + 1 = 5m + 4$ implies $2n + 2 = 5m + 5 = 5(m + 1)$, and so 2 divides $m + 1$. Therefore, $m + 1 = 2k$ for some $k \in \mathbb{Z}$, which we can write as $m = 2k - 1$. This means that $x = 5m + 4 = 5(2k - 1) + 4 = 10k - 5 + 4 = 10k - 1 + (10 - 10) = 10k - 10 + 9 = 10(k - 1) + 9$. Since $(k - 1) \in$, $x \in \{10k + 9 : k \in \mathbb{Z}\}$. Thus we have established that $(\{2n + 1 : n \in \mathbb{Z}\} \cap \{5m + 4 : m \in \mathbb{Z}\}) \subseteq \{10k + 9 : k \in \mathbb{Z}\}$.

4. (a) **Proposition:** For all sets A and B, $A \cap B \subseteq B$
Proof. Let A and B be given. Let $x \in A \cap B$ be given. This means that $x \in A$ and $x \in B$. So $x \in B$. Therefore $A \cap B \subseteq B$.

(b) **Proposition:** For all sets A and B, $B \subseteq A \cup B$
Proof. Let A and B be given. Let $x \in B$ be given. It follows that $x \in A$ or $x \in B$. So $x \in A \cup B$. Therefore $B \subseteq A \cup B$.

(c) **Proposition:** If $A \subseteq B$, then $A \cup B \subseteq B$.
Proof. Let A and B be given such that $A \subseteq B$. Let $x \in A \cup B$ be given. This means that $x \in A$ or $x \in B$, so we can consider two cases:

- CASE 1: If $x \in A$, then since $A \subseteq B$, we can infer that $x \in B$.
- CASE 2: If $x \in B$, then we can also infer that $x \in B$.

So in either case $x \in B$. Therefore $A \cup B \subseteq B$.

5. *Proof.* Let A, B and C be given, and assume that $A \subseteq B$ and $A \subseteq C$. Let $x \in A$ be given. Since $A \subseteq B$, we know that $x \in B$, and since $A \subseteq C$, we know that $x \in C$. We conclude that $x \in (B \cap C)$. Therefore $A \subseteq (B \cap C)$.

6. (a) *Proof.* Let $x \in \{10n - 1 : n \in \mathbb{Z}\}$, so that $x = 10n - 1$ for some $n \in \mathbb{Z}$. We can write, $x = 10n - 1 = 2(5n - 1) + 1$, and since $5n - 1 \in \mathbb{Z}$, $x \in \{2k + 1 : k \in \mathbb{Z}\}$, proving that $\{10n - 1 : n \in \mathbb{Z}\} \subseteq \{2k + 1 : k \in \mathbb{Z}\}$. In a similar manner, we can write $x = 10n - 1 = 5(2n - 1) + 4$, and since $(2n - 1) \in \mathbb{Z}$, $x \in \{5m + 4 : m \in \mathbb{Z}\}$, proving that $\{10n - 1 : n \in \mathbb{Z}\} \subseteq \{5m + 4 : m \in \mathbb{Z}\}$. Now by Exercise 5, we have that $\{10n - 1 : n \in \mathbb{Z}\} \subseteq (\{2k + 1 : k \in \mathbb{Z}\} \cap \{5m + 4 : m \in \mathbb{Z}\})$.

7. *Proof.* Let sets A, B and C be given such that $A \subseteq C$ and $B \subseteq C$. Let $x \in A \cup B$ be given. This means that $x \in A$ or $x \in B$, so we can consider two cases:

- CASE 1: If $x \in A$, then since $A \subseteq C$, we can infer that $x \in C$.
- CASE 2: If $x \in B$, then since $B \subseteq C$, we can infer that $x \in C$.

So in either case $x \in C$. Therefore $A \cup B \subseteq C$.

8. (a) *Proof.* Let $x \in \{4k + 1 : k \in \mathbb{Z}\}$, so that $x = 4k + 1$ for some $k \in \mathbb{Z}$. Since we can write $x = 4k + 1 = 2(2k) + 1$, and $2k \in \mathbb{Z}$, we have that $x \in \{2n + 1 : n \in \mathbb{Z}\}$. Hence, $\{4k + 1 : k \in \mathbb{Z}\} \subseteq \{2n + 1\} : n \in \mathbb{Z}\}$. Now, let $y \in \{4m + 3 : m \in \mathbb{Z}\}$, so that $y = 4m + 3$ for some $m \in \mathbb{Z}$. Since we can write $y = 4m + 3 = 2(2m + 1) + 1$, and $(2m + 1) \in \mathbb{Z}$, we have that $y \in \{2n + 1 : n \in \mathbb{Z}\}$. Hence $\{4m + 3 : m \in \mathbb{Z} \subseteq \{2n + 1 : n \in \mathbb{Z}\}$. By Exercise 7, we now have that $(\{4k + 1 : k \in \mathbb{Z}\} \cup \{4m + 3 : m \in \mathbb{Z}\}) \subseteq \{2n + 1 : n \in \mathbb{Z}\}$.

9. (a) **Proposition:** $A \cup (B \cap C) \subseteq (A \cup B) \cap (A \cup C)$.
Proof. Let A, B and C be given, and let $x \in A \cup (B \cap C)$ be given. This means that $x \in A$ or $x \in B \cap C$, so we can consider two cases.

- If $x \in A$, then we can truthfully say that $x \in A$ or $x \in B$ — that is, $x \in A \cup B$. But we can also truthfully say that $x \in A$ or $x \in C$ — that is, $x \in A \cup C$. Since both of these inferences are valid, in this case we can conclude $x \in (A \cup B) \cap (A \cup B)$.

- If $x \in B \cap C$, then $x \in B$ and $x \in C$, so we can say that $x \in A$ or $x \in B$ — that is, $x \in A \cup B$, and we can also say that $x \in A$ or $x \in C$ — that is, $x \in A \cup C$. Since both of these inferences are valid, in this case we can conclude $x \in (A \cup B) \cap (A \cup B)$.

So in either case $x \in (A \cup B) \cap (A \cup C)$. Therefore $A \cup (B \cap C) \subseteq (A \cup B) \cap (A \cup C)$.

11. (a) **Claim.** If $A \cup B = B$, then $A \cap B = A$.

  Proof. Let sets A and B be given such that $A \cup B = B$. To show that $A \cap B = A$, we must show that $A \cap B \subseteq A$ and $A \subseteq A \cap B$. Proposition 1 establishes that $A \cap B \subseteq A$ is always true, so we only need to establish $A \subseteq A \cap B$ using an element-wise proof.

  Let $x \in A$ be given. Since Practice Problem 1 tells us that $A \subseteq A \cup B$, we can infer that $x \in A \cup B$. Since we are given that $A \cup B = B$, we know that $x \in B$. Since $x \in A$ and $x \in B$, we know that $x \in A \cap B$.

  This establishes that $A \subseteq A \cap B$, and hence we conclude that $A = A \cap B$, as desired.

  (c) **Claim.** If $A \cap B = A$ and $B \cap C = B$, then $A \cap C = A$.

  Proof. Let sets A, B and C be given such that $A \cap B = A$ and $B \cap C = B$. To show that $A \cap C = A$, we must show that $A \cap C \subseteq A$ and $A \subseteq A \cap C$. Proposition 1 establishes that $A \cap C \subseteq A$ is always true, so we only need to establish $A \subseteq A \cap C$ using an element-wise proof.

  Let $x \in A$ be given. Since we are given that $A \cap B = A$, we know that $x \in A \cap B$, which implies that $x \in B$, since $A \cap B \subseteq B$ by Proposition 1. Since we are given that $B \cap C = B$, we know that $x \in B \cap C$, which implies that $x \in C$, since $B \cap C \subseteq C$ by Proposition 1 again. Hence, $x \in A$ **and** $x \in C$, so $x \in A \cap C$. This establishes that $A \subseteq A \cap C$, and hence we conclude that $A = A \cap C$, as desired.

12. (a) *Proof by contradiction.* Assume that **there is** an element $a \in \mathbb{N}$ in both the set $\{2k + 1 : k \in \mathbb{N}\}$ and the set $\{4k : k \in \mathbb{N}\}$. This means that $a = 2K + 1$ for some $K \in \mathbb{N}$ and $a = 4L$ for some $L \in \mathbb{N}$. Combining these facts leads us to conclude that $2K + 1 = 4L$, which implies that

$$2(2L - K) = 1$$

or $2L - K = \frac{1}{2}$. We know (from closure properties of $\mathbb{Z}$) that it is impossible to subtract integers and get a result that is not an integer, so this last statement is absurd. Hence, there is no such number a — that is, $\{2k + 1 : k \in \mathbb{N}\} \cap \{4k : k \in \mathbb{N}\} = \emptyset$

  (c) *Proof by contradiction.* Assume that **there is** an element $(a, b) \in \mathbb{R} \times \mathbb{R}$ in both the set $\{(x, y) \in \mathbb{R} \times \mathbb{R} : x^2 - 2x - 3 = y\}$ and the set $\{(x, y) \in \mathbb{R} \times \mathbb{R} : x - 6 = y\}$. This means that

$$a^2 - 2a - 3 = b = a - 6$$

This in turn implies that $a^2 - 3a + 3 = 0$, which the quadratic formula tells us has no real solutions, a contradiction to the fact that $a \in \mathbb{R}$.

13. (a) **Claim.** If $A \cap B = A$, then $A' \cup B = U$.

  Proof. Let sets A and B of elements in U be given such that $A \cap B = A$. To show that $A' \cup B = U$, we must show that $A' \cup B \subseteq U$ and $U \subseteq A' \cup B$. Since all sets are subsets of the universal set U, we know that $A' \cup B \subseteq U$ is true, so we only need to establish $U \subseteq A' \cup B$ using an element-wise proof.

  Let $x \in U$ be given. By the definition of complement, we know that either $x \in A$ or $x \in A'$, so we consider each case separately to establish that the given x must be in $A' \cup B$.

  - If $x \in A$, then since we are given that $A \cap B = A$, we know that $x \in A \cap B$, from which it follows that $x \in B$. But Practice Problem 1 can be used to establish that $B \subseteq A' \cup B$, so we conclude that $x \in A' \cup B$, as desired.
  - If $x \in A'$, then since $A' \subseteq A' \cup B$ (again by Practice Problem 1), we know that $x \in A' \cup B$, as desired.

  In either case, we see that $x \in A' \cup B$.

  This establishes that $U \subseteq A' \cup B$, and hence we conclude that $A' \cup B = U$.

14.   (a) $(A \cup U) \cap (A \cup \emptyset) = A$

       Proof.

$$
\begin{aligned}
(A \cup U) \cap (A \cup \emptyset) &= A \cap (U \cup \emptyset) && \text{distributive (c)}\\
&= A \cap U && \text{identity (d) or universal bound (i)}\\
&= A && \text{identity (d)}
\end{aligned}
$$

  (c) $A \cup (A' \cap B) = A \cup B$

       Proof.

$$
\begin{aligned}
A \cup (A' \cap B) &= (A \cup A') \cap (A \cup B) && \text{distributive (c)}\\
&= \emptyset \cap (A \cup B) && \text{negation (e)}\\
&= (A \cup B) \cap \emptyset && \text{commutative (a)}\\
&= A \cup B && \text{identity (d)}
\end{aligned}
$$

15.   (a) $(A \cap \emptyset) \cup (A \cap U) = A$

  (c) $A \cap (A' \cup B) = A \cap B$

16.   (a) Prove that $A \cap A = A$.

$$
\begin{aligned}
A &= A \cap U && \text{identity (d)}\\
&= A \cap (A \cup A') && \text{negation (e)}\\
&= (A \cap A) \cup (A \cap A') && \text{distributive (c)}\\
&= (A \cap A) \cup \emptyset && \text{negation (e)}\\
&= A \cap A && \text{identity (d)}
\end{aligned}
$$

  (b) Prove that $A \cup U = U$.

$$
\begin{aligned}
A \cup U &= (A \cup U) \cap U && \text{identity (d)}\\
&= (A \cup U) \cap (A \cup A') && \text{negation (e)}\\
&= A \cup (U \cap A') && \text{distributive (c)}\\
&= A \cup (A' \cap U) && \text{commutative (a)}\\
&= A \cup A' && \text{identity (d)}\\
&= U && \text{negation (e)}
\end{aligned}
$$

  (c) Prove that $A \cap (A \cup B) = A$.

$$
\begin{aligned}
A \cap (A \cup B) &= (A \cup \emptyset) \cap (A \cup B) && \text{identity (d)}\\
&= A \cup (\emptyset \cap B) && \text{distributive (c)}\\
&= A \cup (B \cap \emptyset) && \text{commutative (a)}\\
&= A \cup \emptyset && \text{universal bound (i)}\\
&= A && \text{identity (d)}
\end{aligned}
$$

18.   (a) If $A \cup B = B$, then $A \cap B = A$.

       Proof. Let sets A and B be given such that $A \cup B = B$. Then the following holds true:

$$
\begin{aligned}
A \cap B &= A \cap (A \cup B) && \text{since } B = A \cup B\\
&= A && \text{absorption (j)}
\end{aligned}
$$

  (b) If $A \cap B = A$, then $A \cup B = B$.

       Proof. Let sets A and B be given such that $A \cap B = A$. Then the following holds true:

$$
\begin{aligned}
A \cup B &= (A \cap B) \cup B && \text{since } A = A \cap B\\
&= B \cup (B \cap A) && \text{commutative (a) twice}\\
&= B && \text{absorption (j)}
\end{aligned}
$$

  (c) If $A \cap B = A$, then $A' \cup B = U$.

       Proof. Let sets A and B be given such that $A \cap B = A$. Then the following holds true:

$$
\begin{aligned}
A' \cup B &= (A \cap B)' \cup B && \text{since } A = A \cap B\\
&= (A' \cup B') \cup B && \text{DeMorgan's law (h)}\\
&= A' \cup (B' \cup B) && \text{associative (b)}\\
&= A' \cup (B \cup B') && \text{commutative (a)}\\
&= A' \cup U && \text{negation (e)}\\
&= U && \text{universal bound (i)}
\end{aligned}
$$

21. (a) *Proof.* Suppose C is any set and $A \subseteq B$, and let $(a, c) \in A \times C$. By definition of $A \times C$, we know that $a \in A$ and $c \in C$. Since $A \subseteq B$, it follows that $a \in B$, and so $(a, c) \in B \times C$. This proves that $(A \times C) \subseteq (B \times C)$.

(b) *Proof.* Suppose $A \subseteq B$, and let $C \in \mathcal{P}(A)$. This means that $C \subseteq A$. Since $A \subseteq B$, it follows that $C \subseteq B$, which means that $C \in \mathcal{P}(B)$. This proves that $\mathcal{P}(A) \subseteq \mathcal{P}(B)$.

23. (a) *Proof.* Let $C \in \mathcal{P}(A) \cap \mathcal{P}(B)$ be given. This means that $C \subseteq A$ and $C \subseteq B$. From this it follows that $C \subseteq A \cap B$, and so $C \in \mathcal{P}(A \cap B)$. This establishes that $\mathcal{P}(A) \cap \mathcal{P}(B) \subseteq \mathcal{P}(A \cap B)$.

Now let $C \in \mathcal{P}(A \cap B)$ be given. This means that $C \subseteq (A \cap B)$. Since $(A \cap B) \subseteq A$ and $(A \cap B) \subseteq B$, it follows that $C \subseteq A$ and $C \subseteq B$. That is, $C \in \mathcal{P}(A) \cap \mathcal{P}(B)$. This establishes that $\mathcal{P}(A \cap B) \subseteq \mathcal{P}(A) \cap \mathcal{P}(B)$.

Since we have established each set is a subset of the other, we can conclude that $\mathcal{P}(A) \cap \mathcal{P}(B) = \mathcal{P}(A \cap B)$.

Section 3.4 exercises

1. (a) $(a \cdot b') + a = a$
   (c) $(a' \cdot b)' \cdot (a + b) = a$
   (e) $(a \cdot b)' + (a + b) = 1$

2. (a) $a \cdot (b' + a) = a$
   (c) $(a + b) \cdot (a' \cdot c)\prime = a + (b \cdot c')$
   (e) $(a \cdot b')' = a' + (a \cdot b)$

3. (a) **Claim.** $(a + 1) \cdot (a + 0) = a$

   Proof.

$(a + 1) \cdot (a + 0)$	$= \quad 1 \cdot (a + 0)$	(i) universal bound
	$= \quad (a + 0) \cdot 1$	(a) commutative
	$= \quad a + 0$	(d) identity
	$= \quad a$	(d) identity

   (b) **Claim.** $a \cdot (a' + b) = ab$

   Proof.

$a \cdot (a' + b)$	$= \quad a \cdot a' + a \cdot b$	(c) distributive
	$= \quad 0 + a \cdot b$	(e) negation
	$= \quad a \cdot b + 0$	(a) commutative
	$= \quad a \cdot b$	(d) identity

4. (a) *Proof.* Version 1.

$(a + b) \cdot (b + c)$	$= \quad (a + b) \cdot b + (a + b) \cdot c$	Distributive(c)
	$= \quad b \cdot (b + a) + c \cdot (a + b)$	Commutative(a)
	$= \quad b + c \cdot (a + b)$	Absorption(j)
	$= \quad b + c \cdot a + c \cdot b$	Distributive(c)
	$= \quad b + b \cdot c + a \cdot c$	Commutative(a)
	$= \quad b + a \cdot c$	Absorption(j)
	$= \quad a \cdot c + b$	Commutative(a)

   Version 2.

$a \cdot c + b$	$= \quad b + a \cdot c$	Commutative(a)
	$= \quad (b + a) \cdot (b + c)$	Distributive(c)
	$= \quad (a + b) \cdot (b + c)$	Commutative(a)

   (c) **Claim:** $(a + b)(a'c)' = a + bc'$

   Proof.

$(a + b)(a'c)'$	$= \quad (a + b)(a'' + c')$	(h) DeMorgan law
	$= \quad (a + b)(a + c')$	(f) double negation
	$= \quad a + b \cdot c'$	(c) distributive

41

5. (a) **Claim:** If $a + b = b$, then $a + (ba') = b$

   Proof. Let a, b be given such that $a + b = b$. Then

$$
\begin{aligned}
a + ba' &= a + (a+b) \cdot a' & \text{since } b = a+b \\
&= a + a'(a+b) & \text{Commutative(a)} \\
&= a + a' \cdot a + a' \cdot b & \text{Distributive(c)} \\
&= a + a \cdot a' + b \cdot a' & \text{Commutative(a) twice} \\
&= a + 0 + b \cdot a' & \text{Negation(e)} \\
&= a + b \cdot a' & \text{Identity(d)}
\end{aligned}
$$

6. (a) **Claim.** If $a + b = b$, then $a \cdot b = a$.

   Proof. Let a and b be given such that $a + b = b$. Then

$$
\begin{aligned}
a \cdot b &= a \cdot (a+b) & \text{Since } b = a+b \\
&= a & \text{Absorption(j)}
\end{aligned}
$$

   (c) **Claim.** If $a' + b = 1$, then $a \cdot b' = 0$.

   Proof. Let a and b be given such that $a' + b = 1$. Then

$$
\begin{aligned}
a \cdot b' &= (a')' \cdot b' & \text{Double Negative(f)} \\
&= (a' + b)' & \text{DeMorgan(h)} \\
&= (1)' & \text{Since } a' + b = 1 \\
&= 0 & \text{Complements(k)}
\end{aligned}
$$

7. (a) becomes "If $A \cup B = B$, then $A \cup (B - A) = B$," and

   (b) becomes "If $A \cap B = A$, then $B \cap (B - A)' = A$."

   It is easier to prove with the abstract properties than using definitions of sets.

9. (a) The completed tables are given below:

$\bullet$	1	2	3	5	6	10	15	30
1	1	1	1	1	1	1	1	1
2	1	2	1	1	2	2	1	2
3	1	1	3	1	3	1	3	3
5	1	1	1	5	1	5	5	5
6	1	2	3	1	6	2	3	6
10	1	2	1	5	2	10	5	10
15	1	1	3	5	3	5	15	15
30	1	2	3	5	6	10	15	30

$+$	1	2	3	5	6	10	15	30
1	1	2	3	5	6	10	15	30
2	2	2	6	10	6	10	30	30
3	3	6	3	15	6	30	15	30
5	5	10	15	5	30	10	15	30
6	6	6	6	30	6	30	30	30
10	10	10	30	10	30	10	30	30
15	15	30	15	15	30	30	15	30
30	30	30	30	30	30	30	30	30

   (b) $u = 30$

   (c) $z = 1$

   (d) $1' = 30$, $2' = 15$, $3' = 10$, $5' = 6$, $6' = 5$, $10' = 3$, $15' = 2$, $30' = 1$

11. Properties (a), (c), (d), & (e) (70 plays the role of "1" and 1 plays the role of "0") are all true, so this structure is a Boolean algebra.

12. (a) A check mark indicates those pairs where $L(a, b)$ is true.

$a \backslash b$	1	2	3	5	6	10	15	30
1	✓	✓	✓	✓	✓	✓	✓	✓
2		✓			✓	✓		✓
3			✓		✓		✓	✓
5				✓		✓	✓	✓
6					✓			✓
10						✓		✓
15							✓	✓
30								✓

   (b) "$L(a, b)$" is true precisely when a evenly divides b.

13. (a) **Claim.** $a \cdot a = a$.

Proof.

$$
\begin{aligned}
a &= a \cdot 1 & \text{(d) identity} \\
&= a \cdot (a + a') & \text{(e) negation} \\
&= (a \cdot a) + (a \cdot a') & \text{(c) distributive} \\
&= (a \cdot a) + 0 & \text{(e) negation} \\
&= a \cdot a & \text{(d) identity}
\end{aligned}
$$

Section 3.5 exercises

1. (a) Distributivity: The columns for $a \cdot (b + c)$ and $(a \cdot b) + (a \cdot c)$ are identical.

a	b	c	$b + c$	$a \cdot (b + c)$	$a \cdot b$	$a \cdot c$	$(a \cdot b) + (a \cdot c)$
0	0	0	0	0	0	0	0
0	0	1	1	0	0	0	0
0	1	0	1	0	0	0	0
0	1	1	1	0	0	0	0
1	0	0	0	0	0	0	0
1	0	1	1	1	0	1	1
1	1	0	1	1	1	0	1
1	1	1	1	1	1	1	1

3. (a)

a	b	$a + b$	$b(a + b)$	$a + b(a + b)$
0	0	0	0	0
0	1	1	1	1
1	0	1	0	1
1	1	1	1	1

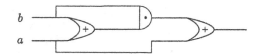

(c)

a	b	c	ab	$ab + c'$	$b' + c$	$(ab + c')(b' + c)$
0	0	0	0	1	1	1
0	0	1	0	0	1	0
0	1	0	0	1	0	0
0	1	1	0	0	1	0
1	0	0	0	1	1	1
1	0	1	0	0	1	0
1	1	0	1	1	0	0
1	1	1	1	1	1	1

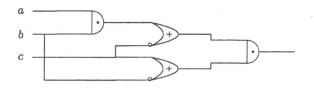

a	b	c	abc	abc'	ab'	abc + abc' + ab'
0	0	0	0	0	0	0
0	0	1	0	0	0	0
0	1	0	0	0	0	0
0	1	1	0	0	0	0
1	0	0	0	0	1	1
1	0	1	0	0	1	1
1	1	0	0	1	0	1
1	1	1	1	0	0	1

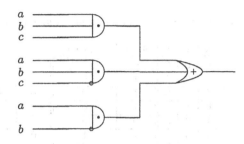

4. (a)
$$
\begin{aligned}
a + b(a + b) &= a + b \cdot a + b \cdot b && \text{(c) Distributive}\\
&= a + b \cdot b && \text{(j) Absorption}\\
&= a + b && \text{(g) Independent}
\end{aligned}
$$

(c)
$$
\begin{aligned}
(ab + c')(b' + c) &= (ab + c')b' + (ab + c') \cdot c && \text{(c) Distributive}\\
&= (abb' + c'b' + abc + c'c) && \text{(c) Distributive}\\
&= a \cdot 0 + c'b' + abc + 0 && \text{(e) Negation}\\
&= 0 + c'b' + abc + 0 && \text{(i) Universal bound}\\
&= c'b' + abc && \text{(d) Identity}
\end{aligned}
$$

(e)
$$
\begin{aligned}
abc + abc' + ab' &= ab(c + c') + ab' && \text{(c) Distributive}\\
&= ab \cdot 1 + ab' && \text{(e) Negative}\\
&= a\bar{b} + ab' && \text{(d) Identity}\\
&= a(b + b') && \text{(c) Distributive}\\
&= a \cdot 1 && \text{(e) Negative}\\
&= a && \text{(d) Identity}
\end{aligned}
$$

5. (a) $ab'c' + abc$

    (b) $a'b'c' + a'b'c + a'bc' + a'bc$

    (c) $a'b'c + a'bc + ab'c + abc$

8. (a) Equivalent

    (c) Equivalent

9. (a) $x'z'$

    (c) z

10. In each of these solutions, to describe the rectangles we number the check marks row by row. For example, in part (a) we number them as (1) $xyzw'$, (2) $xy'zw'$, (3) $xy'z'w'$, and so on.

    (a) Use four rectangles of size 2: check marks 1 and 2 yield xzw'; check marks 3 and 4 yield $xy'z'$; check marks 5 and 6 yield $x'y'z$, and check marks 7 and 8 yield $x'z'w'$. So the solution is $xzw' + xy'z' + x'y'z + x'z'w'$.

    (b) Use a rectangle of size 4 (check marks 1,2,4,6) and two of size 2 (2,3 and 5, 7) to obtain $zw' + xyw + x'yw$.

(c) Use two rectangles of size 4 (1,2,7,10 and 7,8,9,10) and three of size 2 (3,4 and 4,6 and 5,7) to obtain $yw + x'y + xy'w' + y'z'w' + x'zw$.

11. (a) The result $xy + yzw' + x'y'z'$ follows from the following table:

	zw	zw'	$z'w'$	$z'w$
xy	✓	✓	✓	✓
xy'				
$x'y'$			✓	✓
$x'y$		✓		

(c) The result $z + x'y + xy'$ follows from the following table:

	yz	yz'	$y'z'$	$y'z$
x	✓		✓	✓
x'	✓	✓		✓

12. (a) The following table results in the simplified expression $x'y + zw' + xz'w + xy'z$ (or in place of $xy'z$ we could have $xy'w$):

	zw	zw'	$z'w'$	$z'w$
xy		✓		✓
xy'	✓	✓		✓
$x'y'$		✓		
$x'y$	✓	✓	✓	✓

(b) The result $yz + yw' + zw' + xy'z'$ follows from the following table:

	zw	zw'	$z'w'$	$z'w$
xy	✓	✓	✓	
xy'		✓	✓	✓
$x'y'$		✓		
$x'y$	✓	✓	✓	

(c) The result $yw + x'y + xy'w' + y'zw'$ follows from the following table:

	zw	zw'	$z'w'$	$z'w$
xy	✓			✓
xy'		✓	✓	
$x'y'$		✓		
$x'y$	✓	✓	✓	✓

13. (a) The result $y + x'$ follows from the following table:

	yz	yz'	$y'z'$	$y'z$
x	✓	✓		
x'	✓	✓	✓	✓

(c) The result $y' + z'w$ follows from the following table:

	zw	zw'	$z'w'$	$z'w$
x				✓
x'	✓	✓	✓	✓

Section 4.1 exercises

1. (a)

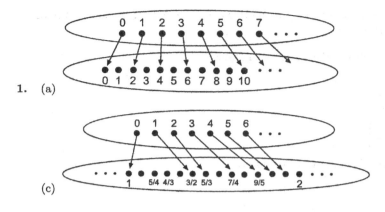

(c)

2. (a) Domain $\{z \in \mathbb{R} \mid z \neq -1\}$, Codomain $\mathbb{R}$

(c) Domain $\{x \in \mathbb{R} \mid x \geq -\frac{1}{2}\}$, Codomain $\{y \in \mathbb{R} \mid y \geq 0\}$

(e) Domain $\mathbb{R}$, Codomain $\{y \in \mathbb{R} \mid 0 < y \leq 1\}$

3. The complete diagram is shown below:

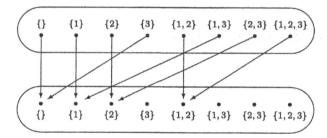

Arrow diagram for f

4. The complete diagram is shown below. Once again, a "loop" at a value indicates that the function maps that value to itself.

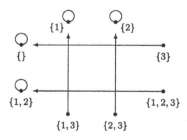

Arrow diagram for f

6. The complete diagram is shown:

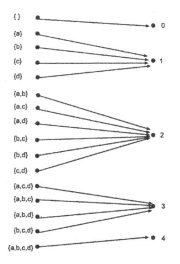

8. The diagrams follow.

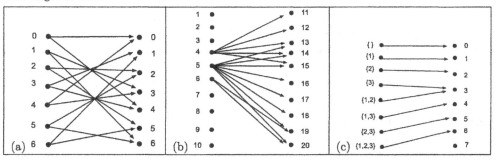

10. (a) Yes, this is a function.

(b) No. The element 1 in the domain is mapped to 2 and 5 in the codomain.

(c) No. The rational number 0.5 can be written as $\frac{1}{2}$ or $\frac{2}{4}$ (as well as many other ways), hence the domain element 0.5 is mapped to 1 and 2 (among other values) in the codomain.

12. (a) and (c) are functions. (b) is not a function because 3 is an element of the domain that is associated with both 6 and 4 in the codomain.

14. (a)

A	$\emptyset$	$\{1\}$	$\{2\}$	$\{3\}$	$\{1,2\}$	$\{1,3\}$	$\{2,3\}$	$\{1,2,3\}$
$C(A)$	$\{1,2,3\}$	$\{2,3\}$	$\{1,3\}$	$\{1,2\}$	$\{3\}$	$\{2\}$	$\{1\}$	$\emptyset$

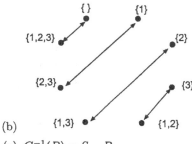

(b)

(c) $C^{-1}(B) = S - B$

16. Claim (i). For all $a, b \in \mathbb{Q}$, if $f(a) = b$, then $g(b) = a$.

Proof. Let $a \in \mathbb{Q}$ and $b \in \mathbb{Q}$ be given such that $f(a) = b$. That is, $3a + 9 = b$. From this it follows that $a = \frac{b-9}{3} = \frac{b}{3} - 3$, and hence $a = g(b)$.

Claim (ii). For all $a, b \in \mathbb{Q}$, if $g(b) = a$, then $f(a) = b$.

Proof. Let $a \in \mathbb{Q}$ and $b \in \mathbb{Q}$ be given such that $g(b) = a$. That is, $\frac{b}{3} - 3 = a$. From this it follows that $b = 3(a + 3) = 3a + 9$, and hence $b = f(a)$.

17. (a) **Claim.** $g(y) = \frac{1}{3}y + 2$ is the inverse of $f(x) = 3x - 6$.

   Proof. Let numbers $a, b \in \mathbb{Q}$ be given such that $f(a) = b$. That is, $3a - 6 = b$, from which it follows algebraically that $3a = b + 6$, or $a = \frac{1}{3}b + 2$. That is, $a = g(b)$.

   Now let numbers $a, b \in \mathbb{Q}$ be given such that $g(b) = a$. That is, $\frac{1}{3}b + 2 = a$, from which it follows algebraically that $\frac{1}{3}b = a - 2$, or $b = 3a - 6$. That is, $b = f(a)$. $\square$

   (b) **Claim.** $g(y) = \frac{1}{2}y - 4$ is the inverse of $f(x) = 2x + 8$.

   Proof. Let numbers $a, b \in \mathbb{Q}$ be given such that $f(a) = b$. That is, $2a + 8 = b$, from which it follows algebraically that $2a = b - 8$, or $a = \frac{1}{2}b - 4$. That is, $a = g(b)$.

   Now let numbers $a, b \in \mathbb{Q}$ be given such that $g(b) = a$. That is, $\frac{1}{2}b - 4 = a$, from which it follows algebraically that $\frac{1}{2}b = a + 4$, or $b = 2a + 8$. That is, $b = f(a)$. $\square$

   (c) **Claim.** $g(y) = 3y + 3$ is the inverse of $f(x) = \frac{1}{3}x - 1$.

   Proof. Let numbers $a, b \in \mathbb{Q}$ be given such that $f(a) = b$. That is, $\frac{1}{3}a - 1 = b$, from which it follows algebraically that $\frac{1}{3}a = b + 1$, or $a = 3b + 3$. That is, $a = g(b)$.

   Now let numbers $a, b \in \mathbb{Q}$ be given such that $g(b) = a$. That is, $3b + 3 = a$, from which it follows algebraically that $3b = a - 3$, or $b = \frac{1}{3}a - 1$. That is, $b = f(a)$. $\square$

   (d) **Claim.** $g(y) = y - 5$ is the inverse of $f(x) = x + 5$.

   Proof. Let numbers $a, b \in \mathbb{Q}$ be given such that $f(a) = b$. That is, $a + 5 = b$, from which it follows algebraically that $a = b - 5$. That is, $a = g(b)$.

   Now let numbers $a, b \in \mathbb{Q}$ be given such that $g(b) = a$. That is, $b - 5 = a$, from which it follows algebraically that $b = a + 5$. That is, $b = f(a)$. $\square$

19. (a) Let $a \in \mathbb{Q}^{\geq 0}$ and $b \in R$ be given such that $f(a) = b$. That is, $\sqrt{a} = b$. From this it follows that $a = b^2$ and hence $a = g(b)$.

   (b) Let $a = 4$ and $b = -2$. Then since $(-2)^2 = 4$, $g(b) = a$ but $f(a) = \sqrt{4} = 2 \neq b$.

   (c) There are g-arrows, like from -2 to 4, that when reversed do not correspond to any f-arrow.

21. The diagrams follow.

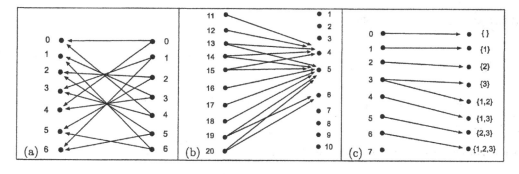

48

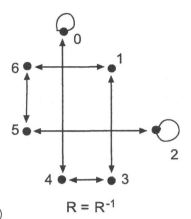

$$R = R^{-1}$$

23. (a)

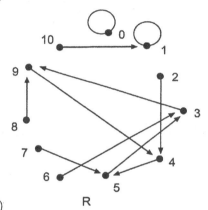

R

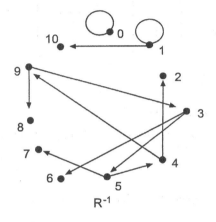

R^{-1}

(b)

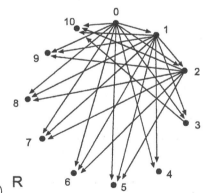

R

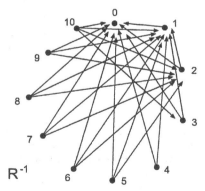

R^{-1}

(c)

25. (a)

26. (a) Yes

Section 4.2 exercises

1. (a) $f(g(z)) = f(z^2 - 1) = 2(z^2 - 1) + 1 = 2z^2 - 1$

   (b) Since $f(g(y)) = 3g(y) - 2$ and $f(g(y)) = 12y + 7$, it follows that $3g(y) - 2 = 12y + 7$, so $g(y) = 4y + 3$.

   (c) Let $g(z) = x$. This means that $2z - 1 = x$, or $z = \frac{1}{2}(x + 1)$. From this, we have $f(x) = f(g(z)) = 6z - 1 = 6\left(\frac{1}{2}(x + 1)\right) - 1$, so $f(x) = 3x + 2$.

3. (a) $f(1) = 11$ and $g(11) = 5$

   (b) The diagrams for f and g appear below:

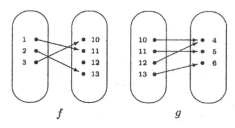

f $\qquad\qquad$ g

   (c) Only $g \circ f$ is defined.

   (d) We can find the rule for function $(g \circ f) : \{1, 2, 3\} \to \{4, 5, 6\}$ by considering each domain element in turn. For example,

$$(g \circ f)(3) = g(f(3)) = g(10) = 4$$

   The complete arrow diagram is given below:

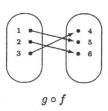

$g \circ f$

5. f is not invertible since $f(0) = f(9) = f(5) = f(4) = 4$, for example, and there is no input that has the output of 9, for example. The diagrams for f, $f \circ f$, and $f \circ f \circ f$ are given below. (In each case, the "loop" at 4 indicates an arrow pointing from 4 to itself.)

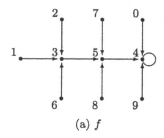
(a) f

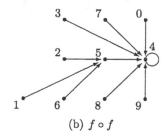
(b) $f \circ f$

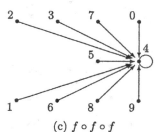
(c) $f \circ f \circ f$

6. The three pictures below provide the missing diagrams. In each case, there is only one possible answer.

50

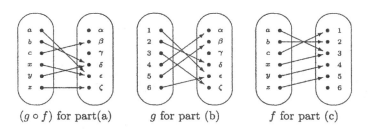

(g ∘ f) for part(a)   g for part (b)   f for part (c)

8. For part (a), $g \circ f$ has domain $\{1, 2, 3, 4\}$, codomain $\{x, y, z\}$, and rule $\{(1, y), (2, x), (3, x), (4, z)\}$. Part (b) is impossible because the codomain of f does not match the domain of g. In part (c), no choice for $f(1)$ can make $(g \circ f)(1) = w$.

10. (a) $f(11011) = 4$, $f(01101) = 3$, and $f(11000) = 2$. This function is not invertible because it is not one-to-one. For example, $f(11011) = f(11110)$.

   (b) $g(0) = 00000$, $g(2) = 11000$, and $g(4) = 11110$. This function is not invertible because it is not onto. For example, there is no value n for which $g(n) = 10101$.

   (c) • $(f \circ g)(2) = f(g(2)) = f(11000) = 2$
       • $(f \circ g)(0) = f(g(0)) = f(00000) = 0$
       • $(g \circ f)(11010) = g(f(11010)) = g(3) = 11100$
       • $(g \circ f)(11100) = g(f(11100)) = g(3) = 11100$

   (d) No. The previous parts show that it is possible to have $f(x) = y$ and $g(y) \neq x$ at the same time — when $x = 11010$ and $y = 3$, for example.

11. *Proof.* Let $f : A \to B$ and $g : B \to A$ be functions that are inverses of each other. From our definition in the previous section, this means that the following equation (∗) is true:

$$\text{For all } a \in A \text{ and } b \in B, \; f(a) = b \text{ if and only if } g(\underline{b}) = \underline{a}$$

We must show both $g \circ f = \iota_A$ and $f \circ g = \iota_B$.

Let $a \in A$ be given, and set $b = f(a)$. In this case, $(g \circ f)(a) = g(f(a)) = g(b)$. The fact that $b = f(a)$ tells us, by equation (∗), that $g(b) = \underline{a}$. Hence, $(g \circ f)(a) = \underline{a}$ for all $a \in A$, proving that $(g \circ f)$ has the same rule as ι_A.

Now let $b \in B$ be given, and set $a = g(b)$. In this case, $(f \circ g)(\underline{b}) = f(a)$. The fact that $a = g(b)$ tells us, by equation (∗), that $f(\underline{a}) = b$. Hence, $(f \circ g)(b) = \underline{b}$ for all $b \in B$, proving that $\underline{(f \circ g)}$ has the same rule as ι_B.

13. Here are the diagrams of R (on the left) and $R \circ R$ (on the right) for each relation.

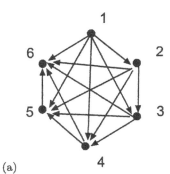

(a)

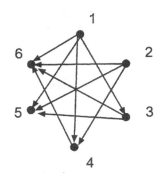

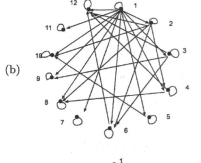

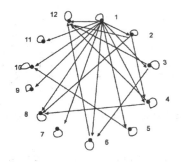

(b)

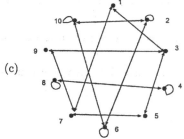

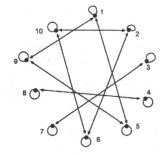

(c)

15. Here is the complete picture:

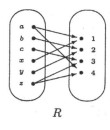

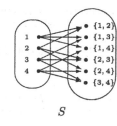

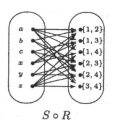

R S $S \circ R$

17. $(x, y) \in R \circ R$ means x beat at least one team that beat y.

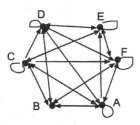

19. Student x is in instructor b's course.

21. $(u, v) \in R$ means that production facility u produces machine part v.

23.

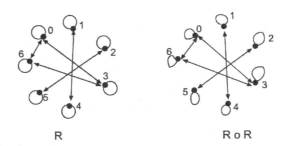

R   R o R

24. (a)

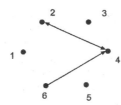

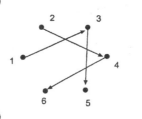

(b)

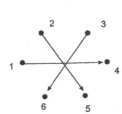

26. (a) $(x,y) \in R_1 \circ R_1^{-1}$ if classes x and y have a student in common.

         $(x,y) \in R_1^{-1} \circ R_1$ if student x and student y have a class in common.

   (c) $(x,y) \in R_3 \circ R_3^{-1}$ if integers x,y are such that $x = \pm y$

         $R_3^{-1} \circ R_3$ relates each nonnegative perfect square to itself

Section 4.3 exercises

1. (a) There is no $x \in R$ for which $f(x) = -4$ since the equation $x^2 + 4x + 1 = -4$ has no real solutions, by the quadratic formula.

   (c) There is no $x \in [1, \infty)$ for which $h(x) = 2$ since the equation $\frac{1}{x+1} = 2$ can only have solution $-\frac{1}{2}$, which is not in $[1, \infty)$.

2. (a) $f(-1) = 1 - 4 + 1 = -2$ and $f(-3) = 9 - 12 + 1 = -2$

   (c) $h(0) = 1$ and $h(-\frac{2}{3}) = 1$

3. (a) One-to-one and onto, and hence invertible

   (c) One-to-one, but not onto because, for example, it is impossible for $\frac{1}{1+x} = 2$ to be true for any value of $x \in \mathbb{R}^+$.

4. (a) f is onto, but it is not one-to-one since, for example, $f(2,2) = 4 = f(3,1)$.

5. (a) c is one-to-one

   (b) c is onto

(c) c is invertible and $c^{-1} = c$

7. *Proof.* We prove that h is one-to-one using the contrapositive of the formal definition of one-to-one. Let $x_1, x_2 \in \mathbb{R}$ be given such that $\underline{h(x_1) = h(x_2)}$. By the definition of h, this means that

$$f(2 \cdot x_1) = f(2 \cdot x_2)$$

However, since f is one-to-one, the only way for $f(2 \cdot x_1) = f(2 \cdot x_2)$ to be true is if $\underline{2 \cdot x_1 = 2 \cdot x_2}$ is true. Dividing by 2 on both sides of this equation, we conclude that $\underline{x_1 = x_2}$, completing the proof.

10. (a) Consider any player of the game. Since the cards were evenly divided, the players all have at least 17 cards (they have 17, 17, and 18). Let A be the set of cards the player has, and let B be the set of possible values (i.e., $\{ace, king, \ldots, 4, 3, 2\}$). Define $f : A \to B$ so that $f(x)$ is the value of the card x. Since A is size 17 or 18, and B is size 13, f is not $1-1$. So there are two cards in A of the same value.

(c) Let $A = \{a, b, c, d, e\}$ be the set of five positive integers, let $B = \{7, 9, 3, 1\}$, and define $f : A \to B$ so that $f(x)$ is the ones' digit of 7^x. Since A has size 5 and B has size 4, f is not $1-1$. Thus there are two numbers y, z in the set A for which 7^y and 7^z have the same ones' digit.

(e) Let A be the set of numbers in the phone number, and let $B = \{0, 1, 2, 3, 4, 5\}$. Define $f : A \to B$ by $f(x) = x \bmod 6$, the remainder when x is divided by 6. Since $n(A) = 7 > 6 = n(B)$, we conclude that f is not $1-1$. For the two numbers x and y for which $f(x) = f(y)$, $x - y$ is divisible by 6.

11. Let $x_1, x_2 \in Z$. Assume that $f(x_1) = f(x_2)$. Then $2x_1 + 3 = 2x_2 + 3$. This implies that $2x_1 = 2x_2$ and it follows that $x_1 = x_2$. Therefore, f is $1-1$.

14. (a) Let $x_1, x_2 \in [0, \infty)$ be given such that $f(x_1) = f(x_2)$. That is, $x_1^2 + 4 = x_2^2 + 4$, so $x_1^2 - x_2^2 = 0$, hence either $x_1 = x_2$ or $x_1 = -x_2$. Since x_1 and x_2 are both from $[0, \infty)$, they must have the same sign, so only the option $x_1 = x_2$ is possible.

(b) Let $y \in [4, \infty)$ be given, and set $x = \sqrt{y-4}$. We can verify with algebra that

$$g(x) = g(\sqrt{y-4})$$
$$= \left(\sqrt{y-4}\right)^2 + 4$$
$$= (y-4) + 4 = y$$

Hence, y is an output of the function g. Therefore g is onto.

(c) $f^{-1}(y) = \sqrt{y-4}$

16. (a) *Proof.* Let $x_1, x_2 \in \mathbb{R}$ be given such that $f(x_1) = f(x_2)$. That is, $x_1^3 - 2 = x_2^3 - 2$, or $x_1^3 = x_2^3$. Taking the cube root of both sides tells us that $x_1 = x_2$, as desired.

(b) *Proof.* Let $y \in \mathbb{R}$ be given, and set $x = \sqrt[3]{y+2}$, a real number. Then $f(x) = x^3 - 2 = (y+2) - 2 = y$. Hence, y is an output of the function f.

(c) $g(y) = \sqrt[3]{y+2}$ is the inverse of f

18. (a) *Proof.* Let $a, b \in (1, \infty)$ be given such that $f(a) = f(b)$. That is, $\frac{a}{a-1} = \frac{b}{b-1}$. Multiplying through by $(a-1)(b-1)$ gives us $ab - a = ab - b$, from which it follows that $a = b$. Therefore, f is one-to-one.

(b) *Proof.* Let $y \in (1, \infty)$ be given, and set $x = \frac{y}{y-1}$. Since $y > y - 1 > 0$, it follows that $\frac{y}{y-1} > 1$. Hence, $x \in (1, \infty)$, and

$$f(x) = \frac{\frac{y}{y-1}}{\frac{y}{y-1} - 1} = \frac{\frac{y}{y-1}}{\frac{y-(y-1)}{y-1}} = \frac{\frac{y}{y-1}}{\frac{1}{y-1}} = y$$

Therefore, y is an output of the function f. Therefore, f is onto.

(c) $f^{-1}(y) = \frac{y}{y-1}$

19. (a) Let $(x_1, y_1), (x_2, y_2) \in A$ be given such that $f(x_1, y_1) = f(x_2, y_2)$. That is, $2^{x_1} \cdot 5^{y_1} = 2^{x_2} \cdot 5^{y_2}$, which implies that $2^{x_1-x_2} = 5^{y_2-y_1}$. The only number that is a power of 2 and a power of 5 is the number 1. Hence, $x_1 - x_2 = y_2 - y_1 = 0$, so $(x_1, y_1) = (x_2, y_2)$. Therefore, f is one-to-one.

(b) Let m be a positive factor of 500. Since $500 = 2^2 \times 5^3$, then m must have the form $2^a \times 5^b$ where $a \in \{0, 1, 2\}$ and $b \in \{0, 1, 2, 3\}$. That is, $m = f(a, b)$. Hence, m is an output of the function f.

(c) Theorem 7

(d) $n(A) = 3 \times 4 = 12$

21. Since $f(abc) = f(acb) = ab$, f is not $1-1$. However, f is onto. Given a word y of length 4 or less over the alphabet $\{a, b\}$, form the word x by putting the letter c at the end of the word y. (For example, if y is $abab$, then x would be $ababc$.) Then x is a word of length 5 or less over the alphabet $\{a, b, c\}$, and $f(x) = y$. The set A is bigger.

24. (a) *Proof.* Let $S \in \mathcal{P}(\{1, 2\})$ be given, and set $T = S \cup \{3\}$. Since $S \subseteq \{1, 2\}$, it follows that $T \subseteq \{1, 2, 3\}$. Moreover, $f(T) = T - \{3\} = S$. Hence, S is an ouput of the function f. Therefore, f is onto.

(b) $f(\{1, 2\}) = f(\{1, 2, 3\}) = \{1, 2\}$

(c) For set S in the codomain $\mathcal{P}(\{1, 2\})$, $f(S \cup \{3\}) = f(S) = S$.

(d) Since every element of the codomain has exactly two arrows pointing to it, we conclude that the domain has twice as many elements as the codomain.

28. The function $f : R^{\geq 0} \to R$ with the rule $f(x) = x$ is one-to-one, and the function $g : R \to R^{\geq 0}$ with the rule $g(y) = 2^y$ is one-to-one. Hence, by Theorem 9, $R^{\geq 0}$ and R have the same size.

29. $B = \{a, c, e\}$

Section 4.4 exercises

1. (a) This relation is transitive, antisymmetric and reflexive.

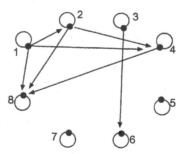

(c) This relation is transitive, but not antisymmetric (since $(2, 8) \in R$ and $(8, 2) \in R$) and not reflexive (since $(5, 5) \notin R$).

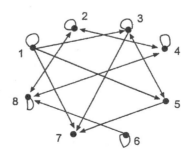

2. (a) Since $a + a = 2a$ is always even, we know that $(a, a) \in R_1$ for all $a \in \mathbb{Z}$. Hence R_1 is reflexive (and hence not irreflexive).

3. (a) Since $(1,3) \in R_1$ and $(3,1) \in R_1$, R_1 is not antisymmetric.

4. (a) **Claim.** R_1 is transitive.

    Proof. Let $(a,b) \in R_1$ and $(b,c) \in R_1$ be given. This means that $a+b$ is even and $b+c$ is even. That is, $a+b = 2K$ and $b+c = 2L$ for some integers K and L. Since

    $$\begin{aligned} a+c &= (a+b)+(b+c)-2b \\ &= 2K+2L-2b \\ &= 2(K+L-b) \end{aligned}$$

    we can see that $a+c$ is even, which means that $(a,c) \in R_1$. Therefore, R_1 is transitive.

5. (a) The blanks might be filled in as follows.
    - $R_1 = \{(a,b) \in A \times A : a \text{ is a } \underline{\text{proper factor of } b}\}$, where $A = \{1, 2, 3, 4, 6, 12\}$
    - $R_2 = \{(a,b) \in B \times B : \underline{a < b}\}$, where $B = \{1, 2, 3, 4, 5\}$

    (b) Neither relation is reflexive. We can extend these relations to the reflexive relations R_1' and R_2':

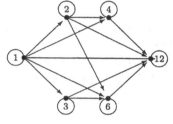 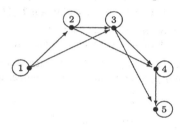

Relation R_1'

Relation R_2'

These can be interpreted as follows: $(x,y) \in R_1'$ means "x divides y"; and $(x,y) \in R_2'$ means "$x \leq y$"

    (c) Both of the relations are transitive.

7. (a) R_1 is irreflexive and antisymmetric. R_1 is not transitive because (my daughter, me) $\in R_1$ and (me, my dad) $\in R_1$ but (my daughter, my dad) $\notin R_1$.

    (b) R_2 is irreflexive, transitive, and antisymmetric.

9. (a) R_1 is reflexive and transitive. R_1 is not antisymmetric because $(\{1,3\},\{2,3\}) \in R_1$ and $(\{2,3\},\{1,3\}) \in R_1$.

    (b) R_2 is antisymmetric and transitive. R_2 is not reflexive because, for example, $(\{1,2\},\{1,2\}) \in R_2$. R_2 is not irreflexive because $(\{1\},\{1\}) \in R_2$.

    (c) R_3 is irreflexive, antisymmetric, and transitive.

11. See Exercise 13 for how the properties of a relation show up in its arrow diagram. This relation is transitive but not antisymmetric, not reflexive and not irreflexive. The arrow diagram is shown below:

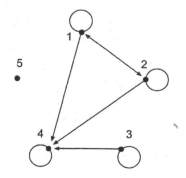

13. (a) There is a loop at every point.

(b) There are no double arrows.

(c) If there's an arrow from a to b and one from b to c, then there must be an arrow from a to c. This must be true even if $a = c$.

15. (a) Relation R_1 is reflexive, antisymmetric and transitive. Its arrow diagram is shown below on the left and its Hasse diagram is shown on the right.

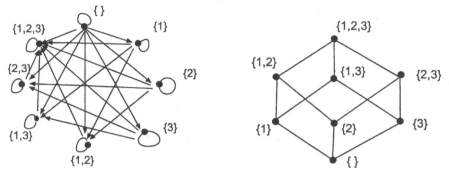

(c) Relation R_3 is not reflexive because $(\{1\}, \{1\}) \notin R_3$. R_3 is not antisymmetric because, for example, $(\{1\}, \{2\}) \in R_3$ and $(\{2\}, \{1\}) \in R_3$. R_3 is not transitive because, for example, $(\{1\}, \{2\}) \in R_3$ and $(\{2\}, \{1,3\}) \in R_3$ but $(\{1\}, \{1,3\}) \notin R_3$. The arrow diagram for R_3 is shown below.

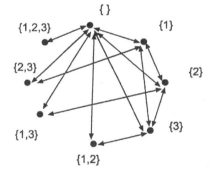

17. (a) If R is reflexive, then R^{-1} is reflexive.

Proof. Let the reflexive relation R on the set A be given. Let $a \in A$ be given. Since R is reflexive, we know $(a, a) \in R$, from which it follows (by reversing the coordinates) that $(a, a) \in R^{-1}$. Hence, R^{-1} is reflexive.

(b) If R is antisymmetric, then R^{-1} is antisymmetric.

Proof. Let the antisymmetric relation R on the set A be given. Let $a, b \in A$ be given such that $(a, b) \in R^{-1}$ and $(b, a) \in R^{-1}$. By definiton of R^{-1}, this means that $(b, a) \in R$ and $(a, b) \in R$, and since R is antisymmetric, it follows that $a = b$. Hence, R^{-1} is antisymmetric.

(c) If R is transitive, then R^{-1} is transitive.

Proof. Let the transitive relation R on the set A be given. Let $a, b, c \in A$ be given such that $(a, b) \in R^{-1}$ and $(b, c) \in R^{-1}$. This means that $(b, a) \in R$ and $(c, b) \in R$, and the transitivity of R tells us that $(c, a) \in R$. From this, it follows that $(a, c) \in R^{-1}$. Hence, R^{-1} is transitive.

20. (a) $R_1 \cup R_2 = \{(0, 1), (1, 2), (0, 3), (1, 4), (0, 0), (1, 1), (2, 1), (2, 2)\}$

(b) $R_1 \cap R_2 = \{(1, 2)\}$

(c) $R_1 - R_2 = \{(0, 1), (0, 3), (1, 4), (0, 0)\}$

21. (a) *Proof.* Let R_1 and R_2 be reflexive relations on the set A. Let $a \in A$ be given. Since R_1 and R_2 are reflexive, we know that $(a, a) \in R_1$ and $(a, a) \in R_2$, from which it follows that $(a, a) \in R_1 \cup R_2$. Hence, $R_1 \cup R_2$ is reflexive.

(c) This statement is false, as we can see from an example of relations on the set $A = \{1, 2, 3\}$. Let $R_1 = \{(1, 2), (1, 3), (2, 3)\}$ and $R_2 = \{(2, 1), (3, 1), (3, 2)\}$. Both of these relations are antisymmetric, while their union

$$R_1 \cup R_2 = \{(1, 2), (1, 3), (2, 3), (2, 1), (3, 1), (3, 2)\}$$

is clearly not antisymmetric, having $(1, 2)$ and $(2, 1)$ both within, for example.

Section 4.5 exercises

1. **Relation R_1** is symmetric and reflexive as indicated by the "double arrows" and the "loops," respectively, in the figure below. Also R_1 is transitive since there are never two sides of a "triangle" without the third side. **Relation R_2** is symmetric for the same reason, but it is not reflexive (since $(1, 1) \notin R_2$) and it is not transitive (since $(7, 4) \in R_2$ and $(4, 8) \in R_2$ but $(7, 8) \notin R_2$). The rightmost figure below shows that **relation R_3** is not symmetric since $(1, 3) \in R_1$ but $(3, 1) \notin R_3$ for example, and R_3 is not reflexive since $(3, 3) \notin R_3$ for example. However, a careful accounting of the arrows of the form $\rightarrow \cdot \rightarrow$ indicates that R_3 is transitive.

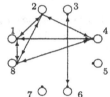
Relation R_1

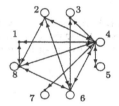
Relation R_2

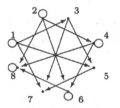
Relation R_3

2. (a) Since addition is commutative, R_1 is symmetric.

(c) Since $(2, 1) \in R_3$ but $(1, 2) \notin R_3$, R_3 is not symmetric.

3. (a) Since addition is commutative, R is symmetric.

(c) Since "equality" of numbers is symmetric, it follows that R is symmetric.

4. (a) This is not a partition since not every element is a set.

(c) This is a partition of A.

6. $\{\{ABCD, BCDA, CDAB, DABC\}, \{ABDC, BDCA, DCAB, CABD\},$
$\{ACBD, CBDA, BDAC, DACB\}, \{ACDB, CDBA, DBAC, BACD\},$
$\{ADBC, DBCA, BCAD, CADB\}, \{ADCB, DCBA, CBAD, BADC\}\}$

8. (a) The relation R_1 is reflexive and transitive, but it is not symmetric since, for example, $(\{1\}, \{1, 2\}) \in R_1$ but $(\{1, 2\}, \{1\}) \notin R_1$. (In fact, R_1 is antisymmetric.) The diagram is shown below.

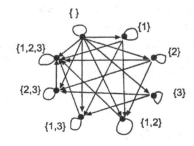

(c) R_3 is symmetric, but it is not reflexive (since $(\{1\},\{1\}) \notin R_3$) and it is not transitive (since $(\{1\},\{3\}) \in R_3$ and $(\{3\},\{1\}) \in R_3$ but $(\{1\},\{1\}) \notin R_3$. The diagram is shown below.

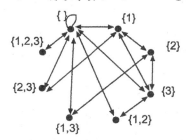

10. (a) *Proof.* Let $a \in A$ be given. Since $a - a = 0$ and 0 is divisible by 3, it follows that $(a, a) \in R$.

(b) *Proof.* Let a and $b \in A$ be given such that $(a, b) \in R$. This means that $x - y = 3 \cdot K$ for some integer K. In this case, $y - x = 3 \cdot (-K)$, so $y - x$ is divisible by 3. This means that $(y, x) \in R$.

(c) *Proof.* Let $a, b, c \in A$ be given such that $(a, b) \in R$ and $(b, c) \in R$. That means $a - b = 3 \cdot K$ and $b - c = 3 \cdot L$ for some integers K and L. In this case,

$$a - c = (a - b) + (b - c)$$
$$= 3K + 3L$$
$$= 3(K + L)$$

Hence, $a - c$ is divisible by 3, so $(a, c) \in R$.

(d) $\{\{0, 3, 6\}, \{1, 4\}, \{2, 5\}\}$

12. In the relation $R \circ R^{-1}$, each person who has a son is related to himself/herself and to his/her "spouse." This relation is not reflexive since, for example, (John, John) $\notin R \circ R^{-1}$. However, $R \circ R^{-1}$ *is* symmetric and transitive.

13. (a) A relation is reflexive if there is a loop at every node.

(b) A relation is symmetric if only double arrows are necessary because all arrows point both ways.

(c) A relation is transitive if every time you see ⟋ , you must also see ⟋ , and every pair of nodes connected by "double arrows" has a loop at each node.

15. (a) *Proof.* Let the symmetric relation R on the set A be given. Let $a, b \in A$ be given such that $(a, b) \in R^{-1}$. This means that $(b, a) \in R$, and since R is symmetric, it follows that $(a, b) \in R$. But this means that $(b, a) \in R^{-1}$. Hence, R^{-1} is symmetric.

(c) *Proof.* Let R be any relation on A. Let $a, b \in A$ be given such that $(a, b) \in R \circ R^{-1}$. This means that, for some $c \in A$, we have $(a, c) \in R^{-1}$ and $(c, b) \in R$. This implies that $(c, a) \in R$ and $(b, c) \in R^{-1}$, from which we can conclude that $(b, a) \in R \circ R^{-1}$. Therefore, $R \circ R^{-1}$ is symmetric.

16. (a) $R_1 \cup R_2 = \{(0,0), (0,1), (0,2), (1,2), (2,1), (3,3), (3,4), (4,3), (4,4)\}$

(b) $R_2 \cap R_2 = \{(1,2), (3,4), (4,3)\}$

(c) $R_2 - R_1 = \{(0,0), (2,1), (3,3), (4,4)\}$

(d) $\{(0,1), (1,2), (0,2), (3,4), (4,3), (3,3), (4,4)\}$

(e) $\{(0,0), (1,1), (1,2), (2,1), (2,2), (3,3), (3,4), (4,3), (4,4)\}$

18. (a) If $R_1 = \{(1,1), (1,2), (2,2), (3,3), (4,4), (5,5)\}$ and $R_2 = \{(1,2), (2,1)\}$, then $R_1 \cap R_2 = \{(1,2)\}$ is neither reflexive nor symmetric.

(b) If $R_1 = \{(1,1), (2,2), (2,3), (3,3), (4,4), (5,5)\}$ and $R_2 = \{(1,2), (2,1)\}$, then

$$R_1 \cup R_2 = \{(1,1), (1,2), (2,1), (2,2), (2,3), (3,3), (4,4), (5,5)\}$$

is reflexive but not symmetric since $(3,2) \notin R_1 \cup R_2$, for example.

(c) If $R_1 = \{(1,1),(1,2),(2,1),(2,2)\}$ and $R_2 = \{(1,1),(1,3),(3,1),(3,3)\}$, then

$$R_1 \cup R_2 = \{(1,1),(1,2),(1,3),(2,1),(2,2),(3,1),(3,3)\}$$

is not transitive since $(2,3) \notin (R_1 \cup R_2)$, for example.

19. (a) $R \cup R^{-1} = \{(0,1),(1,0),(0,0),(1,2),(2,1),(0,3),(3,0),(2,2)\}$

(b) Yes, $R \cup R^{-1}$ is always symmetric.

(c) If R is symmetric, then by Exercise 21, $R = R^{-1}$, so $R \cup R^{-1} = R$.

22. Suppose R is a relation on the set A with the property that $R = R^{-1}$.

Proof that R is symmetric. Let $(a,b) \in R$ be given. By definition of R^{-1}, we know that $(b,a) \in R^{-1}$. Since $R = R^{-1}$, this implies that $(b,a) \in R$. Hence R is symmetric.

Section 4.6 exercises

1. The two columns at the right are identical as predicted by the example.

x	$-x$	$\lfloor -x \rfloor$	$-\lfloor -x \rfloor$	$\lceil x \rceil$
0.5	-0.5	-1	1	1
10.4	-10.4	-11	11	11
-5.3	5.3	5	-5	-5
4	-4	-4	4	4
-17	17	17	-17	-17

2. (a) False. A counterexample is $x = 2.3$, $y = 2.7$.

(c) False. A counterexample is $x = 1.5$.

4. Since $\log_{10} 3^{100} = 100 \cdot \log_{10} 3 \approx 47.7$, we can tell that 3^{100} has 48 digits.

6. Since $\log_{10} 9^{1000} = 1000 \cdot \log_{10}(9) = 2000 \cdot \log_{10}(3) \approx 954.2$, we can tell that 9^{1000} has 955 digits.

8. $a_0 = 1, a_1 = 2^{a_0} = 2, a_2 = 2^{a_1} = 2^2 = 4, a_3 = 2^{a_2} = 2^4 = 16, a_4 = 2^{a_3} = 2^{16} = 65,536$. Since $\log_{10}(a_5) = a_4 \cdot \log_{10}(2) \approx 19728.3$, we know that a_5 has $19\,729$ digits.

10. We have

$$\frac{\log_{10}(3^n)}{\log_{10}(2^n)} = \frac{n \cdot \log_{10}(3)}{n \cdot \log_{10}(2)} = \frac{\log_{10}(3)}{\log_{10}(2)} \approx 1.58$$

so the number of digits in 3^n is roughly 1.6 times the number of digits in 2^n.

11. Let $a = \log_2 10$ and $b = \log_{10} 2$. This means $2^a = 10$ and $10^b = 2$. So $(2^a)^b = 10^b = 2$, which means $a \cdot b = 1$. Hence, $a = \frac{1}{b}$, as desired.

13. In each case, we use Proposition 3 directly.

(a) $\lfloor 2000/3 \rfloor \lfloor 666.666 \ldots \rfloor = 666$.

(b) $\lfloor 2000/5 \rfloor \lfloor 400 \rfloor = 400$.

(c) $\lfloor 2000/7 \rfloor \lfloor 285.71 \ldots \rfloor = 285$.

15. In each case, we use the principle of inclusion/exclusion from Section 3.1. In case this is a little rusty, we include the basic reasoning in our answers as a reminder.

(a) The multiples of 3 plus the multiples of 5 minus the multiples of 15 (which otherwise would be counted twice) gives our desired result. Using Proposition 3, we have

$$\begin{aligned} \lfloor 2000/3 \rfloor + \lfloor 2000/5 \rfloor - \lfloor 2000/15 \rfloor &= 666 + 400 - 133 \\ &= 933 \end{aligned}$$

(c) The multiples of 2 plus the multiples of 5 minus the multiples of 10 (which otherwise would be counted twice) gives our desired result. Using Proposition 3, we have

$$\lfloor 2000/2 \rfloor + \lfloor 2000/5 \rfloor - \lfloor 2000/10 \rfloor \;=\; 1000 + 400 - 200$$
$$=\; 1200$$

(e) We know from part (a) that there are 933 multiples of 3 or 5. The number of these that are even is just the multiples of 6 or 10 in the same range, which is

$$\lfloor 2000/6 \rfloor + \lfloor 2000/10 \rfloor - \lfloor 2000/30 \rfloor = 333 + 200 - 66 = 467$$

Proposition 3 tells us there are 1000 even numbers in this range, so the number of numbers that are multiples of 2 or 3 or 5 is $933 + 1\,000 - 467 = 1\,466$.

Alternatively, we can directly use the "three set" version of the inclusion/exclusion principle as follows:

$$\lfloor 2000/2 \rfloor + \lfloor 2000/3 \rfloor + \lfloor 2000/5 \rfloor \;-\; \lfloor 2000/6 \rfloor - \lfloor 2000/10 \rfloor - \lfloor 2000/15 \rfloor + \lfloor 2000/30 \rfloor$$
$$=\; 1000 + 666 + 400 - 333 - 200 - 133 + 66$$
$$=\; 1\,466$$

18. Since $100! = 100 \cdot 99!$, $then\, 100!$ is simply $99!$ with 2 more zeros on the right.

19. $\log_{10}(100!) = \sum_{i=1}^{100} \log_{10}(i) \approx 157.97$, so $100!$ has 158 digits.

22. $\lfloor \frac{100}{5} \rfloor + \lfloor \frac{100}{25} \rfloor = 20 + 4 = 24$

24. $\lfloor \frac{1992}{3} \rfloor + \lfloor \frac{1992}{9} \rfloor + \lfloor \frac{1992}{27} \rfloor + \lfloor \frac{1992}{81} \rfloor + \lfloor \frac{1992}{243} \rfloor + \lfloor \frac{1992}{729} \rfloor = 664 + 221 + 73 + 24 + 8 + 2 = 992$

27. (a) $\lfloor \frac{500}{2} \rfloor + \lfloor \frac{500}{4} \rfloor + \lfloor \frac{500}{8} \rfloor + \lfloor \frac{500}{16} \rfloor + \lfloor \frac{500}{32} \rfloor + \lfloor \frac{500}{64} \rfloor + \lfloor \frac{500}{128} \rfloor + \lfloor \frac{500}{256} \rfloor =$
$250 + 125 + 62 + 31 + 15 + 7 + 3 + 1 = 494$

28. (a) $\log_2(500!) = \sum_{i=1}^{500} \log_2(i) \approx 3767.4$, so $500!$ requires $3\,768$ binary digits.

29. Since $\lfloor \sqrt{1361} \rfloor = 36$, so only $2, 3, 5, 7, 9, 11, 13, 17, 19, 23, 29,$ and $\;31$ need to be checked.

31. (a) True. $(xy)^z = x^z y^z$
   (c) True. $(x^y)^z = x^{yz}$

32. (a) $Pow(x, Sum(y, z)) = Prod(Pow(x, y), Pow(x, z))$

33. (a) False. $5 \neq \lfloor \frac{5}{2} \rfloor \cdot 2$
   (c) False. $(5 + 3) \bmod 2 \neq (5 \bmod 2) + (3 \bmod 2)$

Section 4.7 exercises

1. (a) $f(1) = O + N + E = 15 + 14 + 5 = 34.$ $f(1) = T + W + O = 96.$

(b) For example, 1 is not in the range. For $f(n)$ to be 1, the word for n would have to be spelled "A," and no number word has that spelling.

(c) Since

$$f(255) \;=\; T + W + O + H + U + N + D + R + E + D + F + I + F + T + Y + F + I + V + E$$
$$=\; 20 + 23 + 15 + 8 + 21 + 14 + 4 + 18 + 5 + 4 + 6 + 9 + 6 + 20 + 25 + 6 + 9 + 22 + 5$$
$$=\; 240$$

and $f(240) = 216$, this completes the cycle.

(d) It takes too long to compute f for one thing, but more importantly f has no fixed point so you cannot know the value at any time.

2. (a) If a card is in position x, then after the deal into three face up piles, the card will be in position $\lceil x/3 \rceil$ of its column. To see this, first observe that 2 out of every 3 cards of the $x-1$ cards that were previously above the given card are now in the other two piles. You cannot have dealt a fraction of a card into one of the other piles, so we have to use the ceiling function to give this value exactly as $\lceil 2/3 \cdot (x-1) \rceil$. Hence the $(x-1) - \lceil 2/3 \cdot (x-1) \rceil$ remaining cards are still above our specified card which means it is in position

$$(x-1) - \lceil 2/3 \cdot (x-1) \rceil + 1, \text{ or}$$
$$x - \lceil 2/3 \cdot (x-1) \rceil$$

of its pile. It remains to be shown that this value is the same as $\lceil x/3 \rceil$ for all x. This can be done by cases on the remainder when x is divided by 3 since in each case we know how the floor and ceiling functions will be evaluated;

- **Case 1.** x is divisible by 3. In this case $\lceil x/3 \rceil = \frac{1}{3}x$, and $x - \lceil 2/3 \cdot (x-1) \rceil = x - \frac{2}{3}x = \frac{1}{3}x$, which is the same thing.
- **Case 2.** $x+1$ is divisible by 3. In this case $\lceil x/3 \rceil = \frac{1}{3}x+\frac{1}{3}$, and $x - \lceil 2/3 \cdot (x-1) \rceil = x - 2\frac{x+1}{3} = \frac{1}{3}x + \frac{1}{3}$, which is the same thing.
- **Case 3.** $x+2$ is divisible by 3. In this case $\lceil x/3 \rceil = \frac{1}{3}x+\frac{2}{3}$, and $x - \lceil 2/3 \cdot (x-1) \rceil = x - 2\frac{x-1}{3} = \frac{1}{3}x + \frac{2}{3}$, which is the same thing.

This means that a card in position x before the deal is in position $11 + \lceil x/3 \rceil$ after the deal.

3. The only two cycles are $1, 6, 3, 8, 4, 2, 1$ and $5, 10, 5$. Formally we prove the following by induction: For every $n \geq 1$, the iterated function sequence for g starting with n leads to one of these two cycles.

Proof. Let $P(n)$ be the statement, "The iterated function sequence for g starting with n leads to one of these two cycles." The numbers in the two given cycles illustrate that $P(1), P(2), \ldots, P(6)$ are all true. Let $m \geq 7$ be given such that the statement $P(m)$ is the first one not yet checked. We argue in two cases depending on whether m is even or odd.

Case 1: If m is even, then $f(m) = m/2$, and the iterated function sequence (i.f.s.) starting with m consists of m followed by the IFS starting with $m/2$. Since $m/2 < m$, we know statement $P(m/2)$ has already been checked to be true, so the iterated function sequence starting with $m/2$ leads to one of the two given cycles.

Case 2: If m is odd, then $f(m) = m + 5$ and $f(f(m)) = \frac{m+5}{2}$, and the iterated function sequence (i.f.s.) starting with m consists of m followed by $m + 5$ followed by the IFS starting with $\frac{m+5}{2}$. Since $\frac{m+5}{2} < m$ (since $m > 5$) we know statement $P\left(\frac{m+5}{2}\right)$ has already been checked to be true, so the iterated function sequence starting with $\frac{m+5}{2}$ leads to one of the two given cycles.

5. (a) In an Excel spreadsheet, we can place an initial value in cell A1 and then the formula

=IF(MOD(A1,3)=0,2*A1/3,IF(MOD(A1,3)=1,(4*A1-1)/3,(4*A1+1)/3))

in cell A2 and filled down through cell A20. We can try for all initial values between 1 and 20. After examining the first few terms of these, we conclude that there is a fixed point at 1 and cycles $2 \to 3 \to 2$ and $4 \to 5 \to 7 \to 9 \to 6 \to 4$.

6. In order for $f(x) = x$, the equation $x^2 - 1 = x$ must be satisfied. The quadratic formula tells us that this is true when

$$x = \frac{1 \pm \sqrt{5}}{2}$$

8. In order for $f(f(x)) = x$, the equation $x^4 - 2x^2 = x$ must be satisfied. Factoring the fourth degree polynomial $x^4 - 2x^2 - x = x(x-1)(x^2 - x - 1)$, we find the solutions

$$x = 0, 1, \frac{1 \pm \sqrt{5}}{2}$$

From Exercise 6, the latter two roots actually indicate cycles of period one, so the cycle $1, 0, 1, 0, 1, 0, \ldots$ is the only one with period two.

10. A one-cycle is a number x for which $f(x) = x$. That is, $x^2 - \frac{1}{2} = x$, or $x^2 - x - \frac{1}{2} = 0$. We can use the quadratic formula to find that the two one-cycles occur at the numbers $\frac{1}{2} + \frac{1}{2}\sqrt{3}$ and $\frac{1}{2} - \frac{1}{2}\sqrt{3}$. A two-cycle involves a number x for which $f(f(x)) = x$ but which is not a one-cycle. That is we need to solve $(x^2 - \frac{1}{2})^2 - \frac{1}{2} = x$, or $x^4 - x^2 - x - \frac{1}{4} = 0$, without getting again the solutions to $x^2 - x - \frac{1}{2} = 0$ that we got in the previous part. We can do this by factoring out the polynomial $x^2 - x - \frac{1}{2}$ from the polynomial $x^4 - x^2 - x - \frac{1}{4}$ like this:

$$x^4 - x^2 - x - \frac{1}{4} = \left(x^2 + x + \frac{1}{2}\right)\left(x^2 - x - \frac{1}{2}\right)$$

and use the quadratic formula to find that there are no real solutions to $x^2 + x + \frac{1}{2} = 0$. Hence there are no two-cycles for this iterated function system.

13. Remember that the letter z is a fixed positive number throughout the process. Solving the equation $g(x) = x$ is the same as solving the equation

$$\frac{x^2 - z}{2x} = 0$$

for which the solutions are $x = \pm\sqrt{z}$. These represent the two cycles of period one. Solving the equation $g(g(x)) = x$ is the same as solving the equation

$$\frac{\left(z + 3x^2\right)\left(-z + x^2\right)}{x\left(x^2 + z\right)} = 0$$

for which the solutions are $x = \pm\sqrt{z}, \pm\sqrt{-z/3}$. The first two of these values represent the two cycles of period one, and the other two values are not real numbers since they involve the square root of a negative number. Hence we conclude that **there are no cycles with period two**.

15. A cycle of period 1 will come from a value of x for which $f(x) = x$. If $x \le 1/2$, this will mean that $2x = x$, which can only happen when $x = 0$. On the other hand, if $x > 1/2$, this will mean that $2 - 2x = x$, which can only happen when $x = 2/3$.

A cycle of period 2 will come from a value of x for which $f(f(x)) = x$. There are four cases to consider:

- If $x \le 1/2$ and $f(x) \le 1/2$, then $f(f(x)) = 4x$, and only $x = 0$ makes $f(f(x)) = x$.
- If $x \le 1/2$ and $f(x) > 1/2$, then $f(f(x)) = 2 - 2f(x) = 2 - 4x$, and $x = 2/5$ makes $f(f(x)) = x$.
- If $x > 1/2$ and $f(x) \le 1/2$, then $f(f(x)) = 2f(x) = 2(2 - 2x)$, and $x = 4/5$ makes $f(f(x)) = x$.
- If $x > 1/2$ and $f(x) > 1/2$, then $f(f(x)) = 2 - 2f(x) = 2 - 2(2 - 2x) = 4x - 2$, and $x = 2/3$ makes $f(f(x)) = x$.

Hence, the only cycle of period 2 is the cycle $2/5, 4/5, 2/5, 4/5, \ldots$.

17. Regardless of the starting value (between 0 and 1), the iterated function sequence in the spreadsheet will eventually be 0 even when we should see the the cycles discussed in Exercise 16. Since computers typically use "base two" representation of numbers, the values of these sequences as fractions with odd denominators cannot be represented exactly. Hence even though we set an initial value like 2/3 in cell A1, the computer stores something very close to but not exactly equal to 2/3 in its memory. This very small initial difference gradually changes the very nature of the sequence as it converges to 0 instead of staying in a cycle.

19. (a) *Proof.* Let $n \ge 2$ be given. Since $F_n = F_{n-1} + F_{n-2}$, it follows (by dividing through by F_{n-1}) that

$$\frac{F_n}{F_{n-1}} = 1 + \frac{F_{n-2}}{F_{n-1}}$$

By definition of r_n, this is the same thing as $r_n = 1 + \frac{1}{r_{n-1}}$.

(b) The function with the rule $g(x) = 1 + \frac{1}{x}$ will create the iterated function sequence with $r_n = g(r_{n-1}) = 1 + \frac{1}{r_{n-1}}$.

(c) A cycle with period 1 comes from a value of x for which $x = g(x)$. With the definition of g given in the previous part, this will be a value of x for which $x = 1 + \frac{1}{x}$. This is the same as satisfying the equation $x^2 - x - 1 = 0$, which by the quadratic formula is true of $x = \frac{1 \pm \sqrt{5}}{2}$. The values of r_n are the ratio of successive Fibonacci numbers, and these ratios "converge" to $\frac{1+\sqrt{5}}{2}$ as n is increased.

Section 4.8 exercises

1. The completed table is shown below.

Length of list (n)	1	2	3	4	5	6	7	8	9	10
Number of comparisons	1	2	2	3	3	3	3	4	4	4
Value of $\lfloor \log_2 (n) \rfloor$	0	1	1	2	2	2	2	3	3	3

2. (a) For $f(n) = n$, when the input value is doubled, the output value is doubled.

(c) For $f(n) = \frac{1}{5}n + 1$, when the input value is doubled, the output value is roughly doubled.

(e) For $f(n) = 3n^2 + 1$, when the input value is doubled, the output value is roughly multiplied by 4.

(g) For $f(n) = 2^{n+1}$, when the input value is doubled, the output value is roughly squared.

4. (a) $2n + 1 \in \Theta(n)$ because $2 \cdot n \leq 2n + 1 \leq 2.5 \cdot n$ for all $n \geq 2$.

(c) $n + \sqrt{n} \in \Theta(n)$ because $1 \cdot n \leq n + \sqrt{n} \leq 2 \cdot n$ for all $n \geq 1$.

5. (a) $3n - 1 \in O(n)$ because $3n - 1 \leq 3 \cdot n$ for all $n \geq 1$.

(b) $10n + 7 \in O(n^2)$ because $10n + 7 \leq 2 \cdot n^2$ for all $n \geq 6$.

6. (a) $3n - 7 \in \Omega(n)$ because $3n - 7 \geq 2 \cdot n$ for all $n \geq 7$.

7. Taking the ratio a_n/n^2 and simplifying yields $\frac{1}{2} - \frac{1}{2n}$. We show that $\frac{1}{4} \leq \frac{1}{2} - \frac{1}{2n} \leq \frac{1}{2}$ for all $n \geq 2$. Since $n > 0$, then $\frac{1}{2n} > 0$, and hence $\frac{1}{2} - \frac{1}{2n} \leq \frac{1}{2}$. On the other hand, since $n \geq 2$, we know that $2n \geq 4$, and therefore $\frac{1}{2n} \leq \frac{1}{4}$, so $\frac{1}{2} - \frac{1}{2n} \geq \frac{1}{2} - \frac{1}{4} = \frac{1}{4}$. By Proposition 3, it follows that $a_n \in \Theta(n^2)$.

9. Taking the ratio a_n/n^3 and simplifying yields $5 + \frac{4}{n} + \frac{6}{n^2} + \frac{7}{n^3}$. For $n \geq 1$, the fractions $\frac{4}{n}$, $\frac{6}{n^2}$ and $\frac{7}{n^3}$ are less than or equal to 4, 6 and 7, respectively. Thus for $n \geq 1$, we have $5 \leq a_n/n^3 \leq 22$. By Proposition 3, it follows that $a_n \in \Theta(n^3)$.

11. (a) Taking the ratio a_n/n^3 and simplifying yields $5 - \frac{4}{n} - \frac{6}{n^2} - \frac{7}{n^3} = 5 - (\frac{4}{n} + \frac{6}{n^2} + \frac{7}{n^3})$. For $n \geq 12$, each of the fractions $\frac{4}{n}$, $\frac{6}{n^2}$, and $\frac{7}{n^3}$ is less than $\frac{1}{3}$. Thus for $n \geq 12$, we have $4 \leq a_n/n^3 \leq 5$.

(b) Taking the ratio a_n/n^3 and simplifying yields $5 + \frac{4}{n} - \frac{6}{n^2} + \frac{7}{n^3}$. This satisfies the inequality

$$5 - \frac{4}{n} - \frac{6}{n^2} - \frac{7}{n^3} \leq 5 + \frac{4}{n} - \frac{6}{n^2} + \frac{7}{n^3} \leq 5 + \frac{4}{n} + \frac{6}{n^2} + \frac{7}{n^3}$$

From part (a), we know that for $n \geq 12$, $\frac{4}{n} + \frac{6}{n^2} + \frac{7}{n^3} \leq 1$. Hence the inequality,

$$5 - \left(\frac{4}{n} + \frac{6}{n^2} + \frac{7}{n^3} \right) \leq a_n \leq 5 + \left(\frac{4}{n} + \frac{6}{n^2} + \frac{7}{n^3} \right)$$

implies the inequality, $4 \leq a_n \leq 6$. Thus for $n \geq 12$, we have $4 \leq a_n/n^3 \leq 6$.

(c) Taking the ratio a_n/n^3 and simplifying yields $5 - \frac{4}{n} - \frac{6}{n^2} + \frac{7}{n^3}$. This satisfies the inequality

$$5 - \left(\frac{4}{n} + \frac{6}{n^2} + \frac{7}{n^3} \right) \leq 5 - \frac{4}{n} - \frac{6}{n^2} + \frac{7}{n^3} \leq 5 + \left(\frac{4}{n} + \frac{6}{n^2} + \frac{7}{n^3} \right)$$

and the result follows just as in part (b).

13. (a) In this case, $\frac{a_n}{n^2} = n$, so $\frac{a_{11}}{11^2} > 10$, $\frac{a_{21}}{21^2} > 20$, $\frac{a_{51}}{51^2} > 50$, and $\frac{a_{101}}{101^2} > 100$.

(c) In this case, $\frac{a_n}{n^2} = n - 5 - \frac{30}{n^2} > n - 6$ when $n \geq 6$, so $\frac{a_{16}}{16^2} > 10$, $\frac{a_{26}}{26^2} > 20$, $\frac{a_{56}}{56^2} > 50$, and $\frac{a_{106}}{106^2} > 100$.

14. (a) Let $a_n = n^2$. Let positive numbers K and N be given. We must find an integer $n > N$, for which $\frac{a_n}{n} > K$. Since $\frac{a_n}{n} = n$, we may take any integer for n that is larger than both K and N. If we do, we will have $n > N$ and also $\frac{a_n}{n} > K$.

15. Answers can vary in each case.

   (a) For $a_n = 2a_{n-1} + n^3$ and $a_1 = 1$, it appears in the spreadsheet that $\frac{1}{4} \cdot n^3 \leq a_n \leq \frac{1}{2} \cdot n^3$ for all $n \geq 10$.

   (b) For $a_n = a_{n-1} + (\frac{1}{2}n + 17)$ and $a_1 = 10$, it appears in the spreadsheet that $\frac{1}{5} \cdot n^2 \leq a_n \leq \frac{1}{2} \cdot n^2$ for all $n \geq 69$.

   (c) For $a_n = a_{n-1} + \sqrt{n}$ and $a_1 = 3$, it appears in the spreadsheet that $\frac{1}{2} \cdot n^{3/2} \leq a_n \leq \frac{3}{4} \cdot n^{3/2}$ for all $n \geq 13$.

18. (a) *Proof by induction.* We will prove that $a_n \leq 3n^2$ for all $n \geq 1$. Since $a_1 = 1$, the first statement, "$a_1 \leq 3$" is true. Let $m \geq 2$ be given such that the statement "$a_m \leq 3m^2$" is the first one not yet checked to be true. In particular, the statement, "$a_{m-1} \leq 3(m-1)^2$" has already been checked to be true. From this we can infer that

$$
\begin{aligned}
a_m &= a_{m-1} + 3m \\
&\leq 3(m-1)^2 + 3m \\
&= 3m^2 - 3(m-1) \\
&\leq 3m^2 \text{ since } m - 1 \geq 0
\end{aligned}
$$

Hence $a_m \leq 3m^2$, which is precisely statement $P(m)$.

   (d) *Proof by induction.* We will prove that $d_n \geq n$ for all $n \geq 1$. Since $d_1 = 1$, the first statement, "$d_1 \geq 1$" is true. Let $m \geq 2$ be given such that the statement "$d_m \geq m$" is the first one not yet checked to be true. In particular, the statement, "$d_{m-1} \geq m - 1$" has already been checked to be true. From this we can infer that

$$
\begin{aligned}
d_m &= d_{m-1} + \lceil \sqrt{m} \rceil \\
&\geq (m-1) + \lceil \sqrt{m} \rceil \\
&= m + (\lceil \sqrt{m} \rceil - 1) \\
&\geq m \text{ since } \lceil \sqrt{m} \rceil \geq 1
\end{aligned}
$$

Hence $d_m \geq m$, which is precisely statement $P(m)$.

20. (a) Since for all $n \in \mathbb{N}$, $1 \cdot f(n) \leq f(n) \leq 1 \cdot f(n)$, it follows that $f(n) \in \Theta(f(n))$.

   (b) Let functions f and g be given such that $f(n) \in \Theta(g(n))$. This means that there is a natural number M such that for all $n \geq M$, $K \cdot g(n) \leq f(n) \leq L \cdot g(n)$ for some positive real numbers K and L. From this it follows that for all $n \geq M$, $\frac{1}{L} \cdot f(n) \leq g(n) \leq \frac{1}{K} \cdot f(n)$, and so $g(n) \in \Theta(f(n))$.

   (c) Let functions f, g and h be given such that $f(n) \in \Theta(g(n))$ and $g(n) \in \Theta(h(n))$. This means that there is a natural number M_1 such that for all $n \geq M_1$, $K_1 \cdot g(n) \leq f(n) \leq L_1 \cdot g(n)$ for some positive real numbers K_1 and L_1, and there is a natural number M_2 such that for all $n \geq M_2$, $K_2 \cdot h(n) \leq g(n) \leq L_2 \cdot h(n)$ for some positive real numbers K_2 and L_2. We choose M to be the larger of M_1 and M_2. In this case, when $n \geq M$, then

$$
\begin{aligned}
f(n) &\leq L_1 g(n) \text{ since } n \geq M_2 \\
&\leq L_1 L_2 h(n) \text{ since } n \geq M_1
\end{aligned}
$$

and

$$
\begin{aligned}
f(n) &\geq K_1 g(n) \text{ since } n \geq M_2 \\
&\geq K_1 K_2 h(n) \text{ since } n \geq M_1
\end{aligned}
$$

It follows from this that $f(n) \in \Theta(h(n))$.

23. (a) We can show that $3^n \notin O(2^n)$ as follows: Let any $K > 0$ be given, and consider an integer n that is greater than $\log_{3/2} K$. In this case,

$$\frac{3^n}{2^n} = \left(\frac{3}{2}\right)^n > \left(\frac{3}{2}\right)^{\log_{3/2} K} = K$$

Since this can be done for any $K > 0$, this shows (by Proposition 3) that $3^n \notin O(2^n)$.

(c) Take $f(n) = 2n^2$ and $g(n) = n^2$. Clearly $f(n) - g(n) \neq 0$.

25. (Proof by induction on k) The proposition contains the predicate we wish to prove, but it is a bit of a mouthful. Therefore, we will use the notation $W(n)$ to stand for the statement, "Algorithm COUNTERFEIT DETECTOR requires k weighings to find a counterfeit among 2^k coins." This statement is clearly true when $k = 1$ since you need one weighing (i.e., weigh either one and see if it is one ounce or not) when you have two coins. That is, $W(1)$ is true.

Now let $m \geq 2$ be given such that statements $W(1), \ldots, W(m-1)$ have all been checked to be true. To verify $W(m)$ we must envision the algorithm being performed on a collection of 2^m coins, one of which is counterfeit. In the first step of the algorithm, the coins are broken into two groups, each consisting of 2^{m-1} coins, and one of the groups is weighed to determine whether it or the other group contains the counterfeit. After this step, the algorithm operates on this remaining collection of 2^{m-1} coins, and so by the previously checked statement $W(m-1)$, it requires $m-1$ more weighings. Therefore, a total of $1 + (m-1) = m$ weighings are needed to find the counterfeit among the original 2^m coins. This is precisely statement $W(m)$ that we had hoped to verify, completing the induction.

26. The algorithm can be modified so that at each step when there are n coins remaining, we take a group of $\lfloor n/2 \rfloor$ coins to be weighed. If this group weighs exactly $\lfloor n/2 \rfloor$ ounces, then the counterfeit coin is in this group and otherwise the counterfeit is in the remaining $\lceil n/2 \rceil$ coins. In the worst case, the counterfeit coin will always be in the larger of the two groups. For example, if there are 9 coins, then after 1 weighing we will narrow it down to 5 coins, after a second weighing we will narrow it down to 3 coins, after a third weighing we will narrow it down to 2 coins, after which a fourth weighing will find the counterfeit for sure. Formally, if W_n is the (worst case) number of weighings with n coins, this reasoning justifies the recurrence relation $W_n = 1 + W_{\lceil n/2 \rceil}$ with $W_2 = 1$. We can use this recurrence relation to complete the table shown below:

Number of coins (n)	1	2	3	4	5	6	7	8	9	10
Number of weighings	0	1	2	2	3	3	3	3	4	4
Value of $\lceil \log_2(n) \rceil$	0	1	2	2	3	3	3	3	4	4

29. (a) C

(b) B

(c) A

32. (a)

k	1	2	3	4	5	6	7	8	9	10
n	2	4	8	16	32	64	128	256	512	1024
c_n	5	19	65	211	665	2059	6305	19171	58025	175099
$n^{\log_2 3}$	3	9	27	81	243	729	2187	6561	19683	59049
Ratio	1.7	2.1	2.4	2.6	2.7	2.8	2.9	2.9	2.95	2.97

(c)

k	1	2	3	4	5	6	7	8	9	10
n	2	4	8	16	32	64	128	256	512	1024
c_n	8	56	296	1400	6248	26936	113576	471800	1939688	7916216
n^2	4	16	64	256	1024	4096	16384	65536	262144	1048576
Ratio	2	3.5	4.6	5.5	6.1	6.6	6.9	7.2	7.4	7.5

33. (a) *Proof by induction.* When $n = 1$, the inequality is "$b_1 \leq 2$," which is true since we are given that $b_1 = 1$ as part of the recursive description. Let $m \geq 2$ be given such that b_m is the first element of the sequence not yet checked to satisfy the inequality. Then

$$\begin{aligned}
b_m &= b_{\lfloor m/2 \rfloor} + m \\
&\leq 2\lfloor m/2 \rfloor + m \\
&\leq 2(m/2) + m = 2m
\end{aligned}$$

Section 5.1

1. (a) List the two choices of candy bar in alphabetical order: {kk, kl, km, ll, lm, mm}

   (b) Since the first bar is given away, order matters this time, so we get {kk, kl, lk, km, mk, ll, lm, ml, mm}

   (c) Number the shirts from 1 to 9, and list the 3 chosen in numerical order: {123, 124, 125, ..., 358, 359, ...}

   (d) The set has all the entries from the previous problem, plus entries such as 111, 344, 667, etc.

   (e) Number the flavors 1 to 8, and list two choices of flavors in numerical order. Here are a few: {11, 12, 13, ..., 18, 22, 23, ..., 77, 78, 88}

   (f) Assuming you think chocolate on top of vanilla is different from vanilla on top of chocolate, you get all the entries from the previous problem, plus their opposites (for example, 21 along with 12).

3. In Exercise 1, we have (a) unordered list (bag) (b) ordered list (c) set (d) bag (e) bag (f) ordered list

   In Exercise 2, we have (a) ordered list (b) ordered list (c) bag (d) ordered list (e) set (f) permutation

4. (a) One does not care about the order in which the cards for a bridge hand are received, and the thirteen cards are distinct, so this would best be represented as a set.

   (c) This could have more than one answer depending on what you do with the dice roll. If you only care about the sum of the upward faces of the dice, for example, then you would probably represent this as an unordered list. Of course, you could have rolled a red die and a green die, and it may be important to you that the red die's value exceeds the green die's value – in this case you would most likely use an ordered list.

   (e) Again it depends, this time on whether you can order more than one of a topping (like extra cheese, double pepperoni, etc.). If so, you had better use an unordered list of toppings. If not, you can use a set to represent the toppings.

5. (a) 36

   (b) 6; 2; A 7 is more likely.

   (c) 10; $\frac{10}{36} = \frac{5}{18} \approx 28\%$

7. (a) 8

   (b) 3

   (c) 7

   (d) 3

9. (a) 6

   (b) 2

   (c) 2

   (d) For player A to win in 3 games, the series must have 2 A's and 1 B, and one of the A's must come last: _, _, A. We define a correspondence from these to the ordered lists of length 2 containg 1 A, by simply dropping the final A.

11. (a) There are 4 columns, each containing 6 rows, for a total of 24 permutations.

    (b) Each group has size 2 (for example, the entries TAHM and HATM become a group), so there are $24 \div 2 = 12$ groups.

(c) Each group gives one distinct arrangement, so there are 12 distinct arrangements just as there are 12

12. (a) We can use this 3×2 table to list the 6 possibilities:

	black pants	white pants
red shirt	red, black	red, white
green shirt	green, black	green, white
yellow shirt	yellow, black	yellow, white

Another possibility is to use a tree, where the first branch indicates the shirt choice and the second branch the choice of pants. This tree will first branch into three choices each of which will branch into two choices giving a total of $3 \cdot 2 = 6$ total choices.

(c) The easiest way to visualize this is by thinking about a tree. The first branch has 3 options, one for each button color. Each of these has 3 options for the second branch (one for each color). After two branches there are nine possibilities. Each of these has 3 options for the third branch, yielding 27 possibilities. Finally, each of these has 3 options for the fourth branch, yielding 81 security codes.

(e) This table will be identical in structure with that in Example 4, with the columns being the arrangements that begin with H, E, A, and R, respectively. There are 24 entries.

HEAR	EHAR	AHER	RHEA
HERA	EHRA	AHRE	RHAE
HAER	EAHR	AEHR	REHA
HARE	EARE	AERH	REAH
HREA	ERHA	ARHE	RAHE
HRAE	ERAH	AREH	RAEH

13. (a) We can use this 4×3 table to list the 12 possibilities.

	return flight 1	return flight 2	return flight 3
flight 1 out	1,1	1,2	1,3
flight 2 out	2,1	2,2	2,3
flight 3 out	3,1	3,2	3,3
flight 4 out	4,1	4,2	4,3

Another possibility is to use a tree, where the first branch indicates the flight out (4 choices) and and the second branch the return flight (3 choices for each of the out flights).

(c) Imagine listing the arrangements on five pages with the headings G, A, M, E and S. The arrangements on the G page will include all of those that start with the letter G. Since they all start with the same letter, there are as many of them as there are ways to arrange the remaining letters AMES. There are 24 ways to arrange the letters of a four-letter word as we saw with the example of MATH. By this reasoning, there are 24 entries on each of the 5 pages, giving us a total of 120 arrangements.

(e) If we use the letters $W = \{S, M, T, W, R, F, A\}$ for the days of the week ($R =$ Thursday and $A =$ Saturday), then this question is asking for the number of two-element subsets of the set W. If we asked instead for people to also rank their favorite two days in order, we would get the following 7×6 table of possible responses:

SM	ST	SW	SR	SF	SA
MS	MT	MW	MR	MF	MA
⋮	⋮	⋮	⋮	⋮	⋮
AS	AM	AT	AW	AR	AF

For each of these ranked responses there is exactly one other that uses the same two days of the week, so there are $42 \div 2 = 21$ different ways to choose (unranked) two days of the week.

14. (a) Use the following tree to see that the answer is 8.

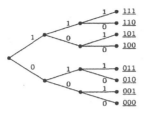

(c) Using the tree for part (b) or by simply making the list $\{1100, 1010, 1001, 0110, 0101, 0011\}$, we can count that there are 6.

(e) To have a binary sequence read the same forward and backwards it must be in the form of A̲ B̲ C̲ B̲ A̲ because the first (A) and last (A) digit are the same and the second (B) and the second to last (B) digit are the same. Hence, once we have chosen the first three digits A̲ B̲ C̲, the rest of the sequence is completely dictated. So the answer to this question is the same as the answer to Exercise 15 (a), which is 8.

16. (a) Following the hint, we construct the following 6×3 table:

Divisible by 4	Divisible by 2 but not 4	Not divisible by 2
4	2	1
$4 \cdot 3 = 12$	$2 \cdot 3 = 6$	$1 \cdot 3 = 3$
$4 \cdot 5 = 20$	$2 \cdot 5 = 10$	$1 \cdot 5 = 5$
$4 \cdot 5^2 = 100$	$2 \cdot 5^2 = 50$	$1 \cdot 5^2 = 25$
$4 \cdot 3 \cdot 5 = 60$	$2 \cdot 3 \cdot 5 = 30$	$1 \cdot 3 \cdot 5 = 15$
$4 \cdot 3 \cdot 5^2 = 300$	$2 \cdot 3 \cdot 5^2 = 150$	$1 \cdot 3 \cdot 5^2 = 75$

Alternatively, the following argument is easier to generalize for the subsequent parts of the problem: Since $300 = 2^2 \cdot 3^1 \cdot 5^2$, its factors all look like $2^- \cdot 3^- \cdot 5^-$ where the first blank is taken from $\{0, 1, 2\}$, the second blank is taken from $\{0, 1\}$ and the third blank is taken from $\{0, 1, 2\}$. There are 18 ways to make these choices.

(b) $600 = \left(2^3\right)\left(3^1\right)\left(5^2\right)$, so every positive factor of 600 must be in the form $(2^w)(3^x)(5^y)$ where w is either 0, 1, 2, 3, x is either 0, 1, and y is either 0, 1, or 2. The 4 possibilities for w, the 2 possibilities for x, and the 3 possibilities for y lead to $4 \cdot 2 \cdot 3 = 24$ different positive factors of 600.

(c) $9576 = (2^3)(3^2)(7^1)(19^1)$, so every positive factor of 9576 must be in the form $(2^w)(3^x)(7^y)(19^z)$ where w is either 0, 1, 2 or 3, x is either 0, 1 or 2, y is either 0 or 1, and z is either 0 or 1. The 4 possibilities for w, the 3 possibilities for x, the 2 possibilities for y and the 2 possibilities for z lead to $4 \cdot 3 \cdot 2 \cdot 2 = \underline{48}$ different positive factors of 9576.

(d) To keep a number square free we must use only exponents of 0 or 1. Hence for 300 there are 8 ways to stick to these choices for each exponent, and for 9576 there are 16 ways to stick to these exponents.

18. (a) Given the set, write a sequence of Is and Os with Os in the positions given by the set.

(b) The completed table is shown below:

OIIOO	OOOII	OOIOI	IOOOI	OIOIO	IOIOO	IOOIO	IIOOO
$\{1,4,5\}$	$\{1,2,3\}$	$\{1,2,4\}$	$\{2,3,4\}$	$\{1,3,5\}$	$\{2,4,5\}$	$\{2,3,5\}$	$\{3,4,5\}$

20. (a) This is simply a matter of changing the A's and B's in an arrangement from (i) into 1's and 0's, respectively. This will change them (one-for-one) into the binary sequences described in (ii).

(b) Items described by (i) look like $\{x, y\}$ where x and y are two different elements from $\{1, 2, 3, 4, 5\}$. Each of these can be "linked" with the set $\{a, b, c\}$ from $\{1, 2, 3, 4, 5\}$ of three entries which are neither x nor y. In this way, each item described by (i) is linked (one-for-one) with an item described by (ii).

(c) For each permutation of all 5 elements we associate the permutation of length 4 that comes from taking the first 4 elements in the same order. For example, to $13ax9$ we associate $13ax$. It is easy to see that this association is one-to-one since the first four entries in the list determine the fifth exactly.

(d) If we represent a sequence of 5 coin tosses as a list of length 5 with entries taken from $\{H, T\}$ then the items in (i) will look like $HHTTH$, $THHHT$, etc. To each of these we can associate the set $\{x, y, z\}$ of the three positions of the "H"s. For example, $HHTTH$ is associated with $\{1, 2, 5\}$, $THHHT$ is associated with $\{2, 3, 4\}$ and so on.

Section 5.2 exercises

1. (a) $6^{10} = 60\,466\,176$; (b) $5^{10} = 9\,765\,625$

3. (a) $26 \cdot 26 \cdot 26 \cdot 10 \cdot 10 \cdot 10 \cdot 10 = 175\,760\,000$

   (b) $(26 \cdot 26 \cdot 26 - 97) \cdot 10 \cdot 10 \cdot 10 \cdot 10 = 174\,790\,000$

5. $3 \cdot 2 \cdot 4 = 24$

6. (a) $P(17, 3) = 17 \cdot 16 \cdot 15 = 4080$

   (b) $16 \cdot 16 \cdot 15 = 3840$

   (c) $16 \cdot 16 \cdot 15 + 1 \cdot 1 \cdot 15 = 3855$

   (d) $1 \cdot 16 \cdot 15 = 240$

8. (a) $8 \cdot 7 \cdot 6 + 12 \cdot 11 \cdot 10 + 6 \cdot 5 \cdot 4 = 1776$

   (b) $26 \cdot 25 \cdot 24 - 20 \cdot 19 \cdot 18 = 8760$

   (c) $8 \cdot 18 \cdot 24 + 18 \cdot 8 \cdot 24 + 8 \cdot 7 \cdot 24 = 8256$

9. (a) $6 \cdot 6 \cdot 6 = 216$

   (b) $6 \cdot 5 \cdot 4 = 120$

11. (a) $P(52, 5) = 52 \cdot 51 \cdot 50 \cdot 49 \cdot 48 = 311\,875\,200$

    (b) $52 \cdot 3 \cdot 50 \cdot 49 \cdot 48 = 18\,345\,600$

    (c) $52 \cdot 12 \cdot 11 \cdot 10 \cdot 9 = 617\,760$

13. (a) $P(8, 8) = 8 \cdot 7 \cdot 6 \cdot 5 \cdot 4 \cdot 3 \cdot 2 \cdot 1 = 40\,320$

    (b) $2 \cdot P(7, 7) = 2 \cdot 7 \cdot 6 \cdot 5 \cdot 4 \cdot 3 \cdot 2 \cdot 1 = 10\,080$

14. (a) $4 \cdot 4 = 16$

    (b) $12 \cdot 32 = 384$

    (c) $1 \cdot 12 + 3 \cdot 13 = 51$

16. (a) $16 \cdot 15 \cdot 14 \cdot 13 = 43\,680$

    (b) $5 \cdot 4 \cdot 14 \cdot 13 = 3\,640$

    (c) $5 \cdot 11 \cdot 14 \cdot 13 + 3 \cdot 13 \cdot 14 \cdot 13 + 8 \cdot 8 \cdot 14 \cdot 13 = 28,756$ or $(16 \cdot 15 \cdot 14 \cdot 13) - (5 \cdot 4 \cdot 14 \cdot 13 + 3 \cdot 2 \cdot 14 \cdot 13 + 8 \cdot 7 \cdot 14 \cdot 13) = 28\,756$

    (d) $5 \cdot 4 \cdot 3 \cdot 2 + 8 \cdot 7 \cdot 6 \cdot 5 = 1\,800$

18. (a) $P(9, 9) = 362\,880$

    (b) $P(5, 5) \cdot P(4, 4) = 2\,880$

20. (a) $12^6 = 2\,985\,984$

    (b) $P(12, 6) = 12 \cdot 11 \cdot 10 \cdot 9 \cdot 8 \cdot 7 = 665\,280$

    (c) $\frac{12^6 - P(12, 6)}{12^6} \approx 77.72\%$ This is a pretty good bet for the speaker.

22. (a) $2^7 = 128$

    (b) $2^7 = 128$

    (c) $2^4 = 16$

(d) $128 + 128 - 16 = 240$

23. (a) $3360 + 1680 + 252 = 5292$

(b) $8568 - (700 + 2520 + 3360 + 1680 + 252) = 56$

(c) $56 + 700 + 2520 = 3276$, or $8568 - 5292 = 3276$

25. (a)

$$
\begin{aligned}
n(A \cup B \cup C) &= n(A) + n(B) + n(C) - n(A \cap B) - n(A \cap C) - n(B \cap C) + n(A \cap B \cap C) \\
&= 75 + 75 + 75 - 24 - 24 - 24 + 6 \\
&= 159
\end{aligned}
$$

(b)

$$
\begin{aligned}
n(A \cup B \cup C) &= n(A) + n(B) + n(C) - n(A \cap B) - n(A \cap C) - n(B \cap C) - n(A \cap B \cap C) \\
&= 4 \cdot 5^3 + 4 \cdot 5^3 + 4 \cdot 5^3 - 12 \cdot 4^2 - 12 \cdot 4^2 - 12 \cdot 4^2 + 12 \cdot 3 \\
&= 960
\end{aligned}
$$

26. (a) $19 \cdot 18 \cdot 17 \cdot 16 \cdot 15 = 1\,395\,360$

(b) $6 \cdot 5 \cdot 4 \cdot 3 \cdot 2 \cdot 1 = 720$

(c) $120 \cdot 119 \cdot 118 \cdot 117 \cdots 106 \cdot 105 \cdot 104 \approx 6.755 \times 10^{34}$

27. (a) $P(9, 4)$

(c) $P(365, 28)$

(e) $P(k, m + 1)$

29. The problem is that, by the time you get to vice-president, there could be either 7 or 6 choices, depending on the sex of the president. You must put the more restrictive steps before the less restrictive steps in your algorithm: "There are 7 choice for the vice-president, 10 for the secretary, and 15 for the president, yielding $7 \cdot 10 \cdot 15 = 1050$ possible results.

31. This algorithm "double-counts" each outcome that has a math major for both of the offices. One way to fix it is to properly apply the Inclusion-Exclusion Principle, by subtracting the overlap. The overlap is given by this algorithm: "8 choices for president, 7 for vice-president, and 24 for secretary," so the total count is $9600 - 8 \cdot 7 \cdot 24 = 8256$. Another possible algorithm counts three non-overlapping sets: (1) president is math major and vice-president is not; (2) vice-president is math major and president is not; (3) both are math majors.[$8 \cdot 18 \cdot 24 + 18 \cdot 8 \cdot 24 + 8 \cdot 7 \cdot 24 = 8256$]. A third algorithm counts the complement (neither are math majors) and subtracts from the overall count [$26 \cdot 25 \cdot 24 - 18 \cdot 17 \cdot 24 = 8256$].

34. We use the following algorithm:

- Choose an odd units' digit. (5 choices)
- Choose a non-zero leftmost digit distinct from the previous choice. (8 choices)
- Choose a tens' digit distinct from the previous choices. (8 choices)
- Choose a hundreds' digit distinct from the previous choices. (7 choices)
- Choose a thousands' digit distinct from the previous choices. (6 choices)

The product rule tells us that $5 \cdot 8 \cdot 8 \cdot 7 \cdot 6 = 13\,440$ numbers will be formed in this way

36. There are $9 \cdot 10 \cdot 10 \cdot 10$ total four-digit numbers and there are $8 \cdot 9 \cdot 9 \cdot 9$ four-digit numbers that do not use a 7 so $9 \cdot 10^3 - 8 \cdot 9^3 = 3\,168$ use the digit 7

37. Such a number can have at most five digits. There are $5 + 5 \cdot 4 + 5 \cdot 4 \cdot 3 + 5 \cdot 4 \cdot 3 \cdot 2 + 5 \cdot 4 \cdot 3 \cdot 2 \cdot 1 = 325$ in all.

39. (a) Choosing the ones' digit first will lead to the total $3 \cdot 5 \cdot 5 \cdot 5 = 375$

(b) Addressing the complementary problem, there are $3 \cdot 4 \cdot 3 \cdot 2$ that have no repeated digits (choosing the ones' digit first and then disallowing repetition of digts), hence there are $3 \cdot 5 \cdot 5 \cdot 5 - 3 \cdot 4 \cdot 3 \cdot 2 = 303$ with some repetition of digits.

(c) We have to deal with this by the possible patterns of the repeated digit. Each case is easy enough to deal with using the product rule.

Case 1. (XXYZ, XYXZ, XYZX, YXXZ, YXZX, YZXX) Choose one of the six patterns, and then choose distinct values for X, Y and Z with the choice of the ones' digit first. This will result in $6 \cdot 3 \cdot 4 \cdot 3 = 216$ numbers.

Case 2. (XXXZ, XXZX, XZXX, ZXXX) Choose one of the four patterns, and then choose distinct values for X and Z with the choice of the ones' digit first. This will result in $4 \cdot 3 \cdot 4 = 48$ numbers.

Case 3. (XXXX) There are 3 such numbers: 1111, 3333, and 5555.

The total number of these is $216 + 48 + 3 = 267$.

41. Form the line according to the following steps:

- Decide on the order of the families in line. (There are $5! = 120$ ways to do this.)
- For each of the five families, decide on the order in which they will stand. (There are $4! = 24$ ways to do this for each of the five families.)

By the product rule, there are $(5!)(4!)^5 = 955\,514\,880$ ways to form the line in this way.

42. There are $\frac{20}{2} = 10$ subsets which is half the number of permutations.

Section 5.3 exercises

1. (a) $ab, ac, ad, ae, ba, bc, bd, be, ca, cb, cd, ce, da, db, dc, de, ea, eb, ec, ed$

(c) 2, 10

2. (a) $5 \cdot 4 \cdot 3 = 60$

(c) $\{ade, aed, dae, dea, ead, eda\}$

(e) $\frac{60}{6} = 10$

3. (a) $\{ACDBFE, CDBFEA, DBFEAC, BFEACD, FEACDB, EACDBF\}$

5. The first step can be completed in $4!/4 = 6$ ways (similar to Example 3), and the second step is really four steps (one for each couple) that each have two ways to be done. The product rule tells us this will result in $6 \cdot 2^4 = 96$ possible seatings.

7. Place the children as follows:

- Take a particular boy (let's say the tallest), and place him anywhere at the table. (1 way to do this)
- Choose a girl to place to his right. (4 ways)
- Choose a different boy to place to her right. (3 ways)
- Choose a different girl to place to his right. (3 ways)
- Choose a different boy to place to her right. (2 ways)
- Choose a different girl to place to his right. (2 ways)
- Choose a different boy to place to her right. (1 way)
- Choose a different girl to place to his right. (1 way)

The product rule tells us there are $4 \cdot 3 \cdot 3 \cdot 2 \cdot 2 \cdot 1 \cdot 1 = 144$ total ways to place the children.

10. By the sum rule, the number of subsets of size 1 or 3 or 5 is $C(6,1) + C(6,3) + C(6,5) = 32$.

12. $P(6,3) = 120$

14. She must choose 3 of the remaining 7, so $C(7,3) = 35$ is the number of ways.

15. (a) $C(21,4) = 5\,985$

   (c) Considering cases based on the number of men being either 1 or 0, we have the sum $C(9,1) \cdot C(12,3) + C(12,4) = 2\,475$

17. (a) $C(12,3) = 220$

   (c) $C(2,2) \cdot C(10,1) = 10$

19. (a) $2^8 = 256$

   (c) From the 8 positions, choose 3 to be M; $C(8,3) = 56$

21. First solution: Divide the list into three disjoint parts: those that have Jack but not Jill; those that have Jill but not Jack; those that have neither, giving $C(1,1) \cdot C(16,4) + C(1,1) \cdot C(16,4) + C(2,0) \cdot C(16,5) = 8008$. Second solution: Solve the complementary problem, giving $C(18,5) - C(2,2) \cdot C(16,3) = 8008$. Third solution: divide the list into two overlapping parts: (i) those that don't have Jack and (ii) those that don't have Jill, and use the sum rule (with overlap), giving $C(1,0) \cdot C(17,5) + C(1,0) \cdot C(17,5) - C(2,0) \cdot C(16,5) = 8008$

23. $C(43,4) \cdot C(47,5) = 189\,303\,411\,990$

24. Note that in this problem, for typographical reasons we use the alternative notation $\binom{n}{k}$ for $C(n,k)$, the number of ways to choose a subset of size k from a set of size n. We can break the committee selections up into cases based on how many married women are chosen:

- **Case 1:** If no married women are used, there are $\binom{23}{4}$ ways to choose the women for the committee and $\binom{47}{5}$ ways to then choose the men. By the product rule there are $\binom{23}{4}\binom{47}{5}$ committees of this type.

- **Case 2:** If one married woman is used, there are $\binom{20}{1}$ ways to choose that woman, then $\binom{23}{3}$ ways to choose the remaining women for the committee, and $\binom{46}{5}$ ways to then choose the men. By the product rule there are $\binom{20}{1}\binom{23}{3}\binom{46}{5}$ committees of this type.

- **Case 3:** If two married women are used, there are $\binom{20}{2}$ ways to choose those women, then $\binom{23}{2}$ ways to choose the remaining women for the committee, and $\binom{45}{5}$ ways to then choose the men. By the product rule there are $\binom{20}{2}\binom{23}{2}\binom{45}{5}$ committees of this type.

- **Case 4:** If three married women are used, there are $\binom{20}{3}$ ways to choose those women, then $\binom{23}{1}$ ways to choose the remaining women for the committee, and $\binom{44}{5}$ ways to then choose the men. By the product rule there are $\binom{20}{3}\binom{23}{1}\binom{44}{5}$ committees of this type.

- **Case 5:** If four married women are used, there are $\binom{20}{4}$ ways to choose those women for the committee, and $\binom{43}{5}$ ways to then choose the men. By the product rule there are $\binom{20}{4}\binom{43}{5}$ committees of this type.

Since the five cases above cover all possible types of committee, the sum rule tells us that the total number of committees is

$$\binom{23}{4}\binom{47}{5} + \binom{20}{1}\binom{23}{3}\binom{46}{5} + \binom{20}{2}\binom{23}{2}\binom{45}{5} +$$
$$+ \binom{20}{3}\binom{23}{1}\binom{44}{5} + \binom{20}{4}\binom{43}{5} = 154\,004\,008\,725$$

25. $C(52,5) - C(4,0) \cdot C(48,5) = 2\,598\,960 - 1\,712\,304 = 886\,656$

27. (a) $C(22,6) = 74\,613$

   (b) $C(10,2) \cdot C(8,2) \cdot C(4,2) = 7560$

29. (a) $C(80, 11) = 10,477\,677\,064\,400$

   (c) About $9\,626\,410$ to 1

32. $C(7, 3) = 35$ is the coefficient of t^6, and the coefficient of t^5 is 0 since all of the powers of t must be even due to the t^2 in the original function.

35. Note that in this problem, for typographical reasons we use the alternative notation $\binom{n}{k}$ for $C(n, k)$, the number of subsets of size k that can be formed from a set of size n.

Proof by induction. Let $P(n)$ be the statement, "$\sum_{k=0}^{n} \binom{n}{k} = 2^n$" It is easy to check that $P(0)$ (which says "$\binom{0}{0} = 2^{0}$") and $P(1)$ (which says "$\binom{1}{0} + \binom{1}{1} = 2^{1}$") are both true. Now let $m \geq 2$ be given such that statements $P(0)$, $P(1)$, ..., $P(m-1)$ have all been checked, and $P(m)$ is the next statement to be checked. So

$$
\begin{aligned}
\sum_{k=0}^{m} \binom{m}{k} &= \binom{m}{0} + \sum_{k=1}^{m-1} \left(\binom{m-1}{k-1} + \binom{m-1}{k} \right) + \binom{m}{m} \\
&= \binom{m-1}{0} + \sum_{k=1}^{m-1} \left(\binom{m-1}{k-1} + \binom{m-1}{k} \right) + \binom{m-1}{m-1} \\
&= \sum_{k=0}^{m-1} \binom{m-1}{k-1} + \sum_{k=0}^{m-1} \binom{m-1}{k} \\
&= 2^{m-1} + 2^{m-1} = 2^m
\end{aligned}
$$

Hence statement $P(m)$ is also true.

39. The 3rd entry of Row 7 is the coefficient of x^3 in the expansion of $(1+x)^7$. On the other hand, $(1+x)^7 = (1+x)^2(1+x)^5$ and

$$(1+x)^2(1+x)^5 = (C(2,0) + C(2,1)x + C(2,2)x^2) \cdot (C(5,0) + C(5,1)x + C(5,2)x^2 + \cdots + C(5,5)x^5)$$

The coefficient of x^3 in the product on the right hand side comes from all ways of multiplying terms from the two polynomials to result in an x^3 term: $C(2,0) \cdot C(5,3)x^3$, $C(2,1)x \cdot C(5,2)x^2$, and $C(2,2)x^2 \cdot C(5,1)x$. Summing these coefficients gives us the original expression under consideration, $C(2,0) \cdot C(5,3) + C(2,1) \cdot C(5,2) + C(2,2) \cdot C(5,1)$.

41. (a) We have worked before (in Chapter 2) with these "geometric" sums of the form $1 + r + r^2 + \cdots + r^n$. The general idea is to use the way we multiply polynomials to see that the following is always true.

$$(1 - r)(1 + r + r^2 + \cdots + r^n) = 1 - r^{n+1}$$

Now putting $r = 1 + x$ in this formula gives us the desired equation.

43. This problem uses the mod 10 arithmetic from Section 2.7. Let A, B, C, D and E be the values of the original cards. The rows formed are

- $[A, B, C, D, E]$
- $[A + B, B + C, C + D, D + E]$
- $[A + 2B + C, B + 2C + D, C + 2D + E]$
- $[A + 3B + 3C + D, B + 3C + 3D + E]$
- $[A + 4B + 6C + 4D + E]$

where each operation uses mod 10 arithmetic. Since $4 \equiv_{10} -6$, we know that

$$A + 4B + 6C + 4D + E \equiv_{10} A + 4B - 4C + 4D + E$$

The right-hand side of this equation is precisely the calculated value $A + E + 4(B + D - C)$ in the magic trick.

74

Section 5.4 exercises

1. (a) $C(8,6) = 28$

   (c) $2^8 - (C(8,0) + C(8,1)) = 247$

3. (a) $C(12,11) \cdot 2 = 24$

   (c) 2^{12}

5. (a) $4^{10} = 1,048,576$

   (c) Place the 0's first, anywhere but in the first position. $C(9,2) \cdot C(8,3) \cdot 3^5 = 489,888$

7. We can choose 3 of the 10 available spaces for the 1's, then 3 of the remaining 7 spaces for the 2's, and finally fill in the four 3's in the only 4 spots left. By the product rule, there are $C(10,3) \cdot C(7,3) \cdot C(4,4) = 4,200$ such numbers.

9. We can find all sequences of length 20 which use each symbol twice taking care not to have a leading 0 digit and hence count those sequences which are not properly written as numbers. We choose 2 of the 19 spaces that do not include the leading digit for the 0's, then 2 of the remaining 18 spaces for the 1's, then 2 of the remaining 16 spaces for the 2's, etc. The product rule tells us that there will be

$$C(19,2)C(18,2)C(16,2)\cdots C(4,2)C(2,2)$$

such numbers.

13. (a) $4! = 24$ weird arrangements look like ALAAMAB once the goofiness is removed.

    (b) $\frac{7!}{4!} = 7 \cdot 6 \cdot 5 = 210$

15. (a) The equations are 2+2+2+2+2+2=12, 6+0+0+0+6+0=12, and 0+0+0+12+0+0=12. The binary sequences are 00100100100100100, 00000011110000001, and 11100000000000011.

17. $C(23,20) = 1\,771$

19. This is the same as the number of arrangements of 11 "0"s and 2 "1"s which is $C(13,2) = 78$.

21. We use cases based on the value of z:

    - If $z = 0$, then the x, y and z are non-negative integers satisfying $x + y + z = 10$. There are $C(12,2) = 66$ such solutions.

    - If $z = 1$, then the x, y and z are non-negative integers satisfying $x+y+z = 8$. There are $C(10,2) = 45$ such solutions.

    - If $z = 2$, then the x, y and z are non-negative integers satisfying $x+y+z = 6$. There are $C(8,2) = 28$ such solutions.

    - If $z = 3$, then the x, y and z are non-negative integers satisfying $x+y+z = 4$. There are $C(6,2) = 15$ such solutions.

    - If $z = 4$, then the x, y and z are non-negative integers satisfying $x + y + z = 2$. There are $C(4,2) = 6$ such solutions.

    - If $z = 5$, then the x, y and z are non-negative integers satisfying $x + y + z = 0$. There is $C(2,2) = 1$ such solution.

    Hence, there are $\sum_{i=0}^{6} C(2i,2) = 161$ solutions in all.

23. $C(10,3) = 120$

25. Following the hint, we count these solutions in four cases:

    Case 1. If $c = 1$, then we are looking for positive integer solutions to the equation $a + b = 8$. Following the reasoning process in Example 8, this is the same as the number of non-negative integer solutions to $a' + b' = 6$, of which there are $C(6 + 2 - 1, 6) = 7$.

Case 2. If $c = 2$, then we are looking for positive integer solutions to the equation $a + b = 6$. This is the same as the number of non-negative integer solutions to $a' + b' = 4$, of which there are $C(4 + 2 - 1, 4) = 5$.

Case 3. If $c = 3$, then we are looking for positive integer solutions to the equation $a + b = 4$. This is the same as the number of non-negative integer solutions to $a' + b' = 2$, of which there are $C(2 + 2 - 1, 2) = 3$.

Case 4. If $c = 4$, then we are looking for positive integer solutions to the equation $a + b = 2$, and clearly $a = b = 1$ is the only one.

There are therefore a total of $7 + 5 + 3 + 1 = 16$ solutions to the original equation.

27. We first give each person a magnet. Now we draw 20 names from the 15 members, with repetitions allowed ($B(15, 20)$). This corresponds to binary sequences with 20 0's and 14 1's, of which there are $C(34, 20) = C(34, 14) = 1,391,975,640$.

29. We first find the number of ways to distribute the apples so that the Democrats and Republicans each get at least one apple by giving them each one and freely distributing the remaining 111 apples among all 100 people. There are $C(111 + 99, 99)$ such distributions. We now count how many of these have all of the Independents receiving too many (in this case, four or more) apples by additionally giving each Independent four apples before freely distributing the remaining 67 apples freely among all 100 people. There are $C(67 + 99, 99)$ such distributions. We conclude that there are $C(210, 99) - C(166, 99)$ ways to distribute the apples in the way allowed.

31. We wish to find the number of integer solutions to the equation

$$x_1 + x_2 + x_3 + x_4 = 12$$

which have each $x_i \geq 1$ from $\{1, 2, \ldots, 6\}$. We first find that there are $C(11, 3)$ solutions with only the restriction that each $x_i \geq 1$. We then subtract those solutions that have any of the $x_i \geq 7$ noting that it is impossible for two of the x_i to be too big in the same solution. There are $C(5, 3)$ solutions in which x_1 is too big, so by symmetry there are $4C(5, 3)$ solutions in which one of the x_i is too big. Hence the number of good solutions is $C(11, 3) - 4C(5, 3) = 125$.

Section 5.5 exercises

1. Let a_n be the number of n-digit numbers which do not use "0" for a digit. Any such n-digit number can be made by choosing a non-zero digit as the leading digit, and then following it with an $(n - 1)$-digit number with the same property. This means that

$$a_n = 9 \cdot a_{n-1}$$

This along with the fact that $a_1 = 9$ completely solves the problem.

2. $d_{1,1} = d_{2,1} = d_{3,1} = d_{4,1} = d_{5,1} = d_{6,1} = 1$ and $d_{n,1} = 0$ for $n \geq 7$

3. *Proof by induction.* From Example 5 we know $T_1 = 0$, $T_2 = 2$ and $T_n = 2 \cdot T_{n-2}$. Let $P(n)$ be the statement, "If n is even, then $T_n = 2^{n/2}$; if n is odd, then $T_n = 0$" for $n \geq 2$. Since $T_1 = 0$ and $T_2 = 2 = 2^{2/2}$ from the example, we know that statements $P(1)$ and $P(2)$ and true. Now suppose we have checked $P(1)$, $P(2)$, $\ldots$, $P(m - 1)$ have all been checked to be true for some integer $m \geq 3$. We have two cases to consider depending on whether m is even or odd.

Case 1. If m is odd, then

$$\begin{aligned} T_m &= 2 \cdot T_{m-2} \text{ from Example 5} \\ &= 2 \cdot 0 \text{ by statement } P(m - 2) \text{ since } m - 2 \text{ is also odd} \\ &= 0 \end{aligned}$$

Case 2. If m is even, then

$$
\begin{aligned}
T_m &= 2 \cdot T_{m-2} \text{ from Example 5} \\
&= 2(2^{(m-2)/2}) \text{ by statement } P(m-2) \text{ since } m-2 \text{ is also even} \\
&= 2(2^{m/2-1}) \\
&= 2^{m/2}
\end{aligned}
$$

In either case, we have confirmed the truth of statement $P(m)$, completing the induction.

5. (a) $P(n,1) = n \cdot P(n-1,0) = n \cdot 1 = n$

(b) $P(n,2) = n \cdot P(n-1,1) = n \cdot (n-1)$ by part (a)

7. Let c_n be the number of ways to cover a $2 \times n$ chessboard with 1×2 dominoes. For example, $c_3 = 3$ since we can cover a 2×3 chessboard in the following ways:

Similarly one can verify that $c_1 = 1$ and $c_2 = 2$. To see the recursive model, just observe that the left side of every covering looks like one of the following:

In the first case, the remaining $2 \times (n-1)$ chessboard must be covered (we know there are c_{n-1} ways to do this), and in the second case, the remaining $2 \times (n-2)$ chessboard must be covered (we know there are c_{n-2} ways to do this). Hence we have $c_n = c_{n-1} + c_{n-2}$.

9. *Proof by induction.* Since $\sum_{k=0}^{1}(-1)^k \frac{1}{k!} = 1 - 1 = 0$ and $d_1 = 0$, the statement is true when $n = 1$. Since $2! \sum_{k=0}^{2}(-1)^k \cdot \frac{1}{k!} = 2(1 - 1 + \frac{1}{2}) = 1$ and $d_2 = 1$, the statement is true when $n = 2$. Let $m \geq 3$ be given such that the first statement that has not yet been checked. Then

$$
\begin{aligned}
d_m &= (m-1) \cdot (d_{m-1} + d_{m-2}) \\
&= (m-1)(m-1)! \cdot \sum_{k=0}^{m-1}(-1)^k \frac{1}{k!} + (m-1) \cdot (m-2)! \sum_{k=0}^{m-2}(-1)^k \cdot \frac{1}{k!} \\
&= m \cdot (m-1)! \sum_{k=0}^{m-1}(-1)^k \cdot \frac{1}{k!} - (m-1)! \sum_{k=0}^{m-1}(-1)^k \frac{1}{k!} + (m-1)! \sum_{k=0}^{m}(-1)^k \frac{1}{k!} \\
&= m! \sum_{k=0}^{m-1}(-1)^k \frac{1}{k!} - (m-1)! \cdot (-1)^{m-1} \cdot \frac{1}{(m-1)!} \\
&= m! \sum_{k=0}^{m-1}(-1)^k \frac{1}{k!} + (-1)^m \\
&= m! \sum_{k=0}^{m}(-1)^k \frac{1}{k!}
\end{aligned}
$$

12. It is straightforward to see that $f_{0,k} = 1$ for all $k \geq 0$, and $f_{n,1} = 1$ for all $n \geq 1$. These will serve as our "initial conditions" for the recursive model. To find a recurrence relation, consider the problem counted by $f_{n,k}$, "How many bags of n pieces of fruit can be filled at a store that carries k types of fruit?" Let's suppose that one type of fruit is apples. We can answer the problem using cases based on the number of apples purchased.

Case 0. If you purchase 0 apples, then there are $f_{n,k-1}$ ways to complete the bag from the $k-1$ remaining types of fruit.

Case 1. If you purchase 1 apple, then there are $f_{n-1,k-1}$ ways to complete the bag from the $k-1$ remaining types of fruit.

Case 2. If you purchase 2 apples, then there are $f_{n-2,k-1}$ ways to complete the bag from the $k-1$ remaining types of fruit.

$$\vdots$$

Case n. If you purchase n apples, then there are $f_{0,k-1}$ ways to complete the bag from the $k-1$ remaining types of fruit.

Since one of these cases must hold for every bag, we have $f_{n,k} = f_{n,k-1} + f_{n-1,k-1} + f_{n-2,k-1} + \cdots + f_{0,k-1}$, or in summation notation,

$$f_{n,k} = \sum_{m=0}^{n} f_{m,k-1}$$

14. Let w_n be the number of positive n-digit numbers with no consecutive 1's. We observe that $w_1 = 9$ and $w_2 = 89$ (any number from 10 to 99 except 11). Assume that we understand these sorts of numbers when they have less than n-digits but we are now asked to count how many are like this with n-digits. To form these we can add $0, 2, 3, 4, \ldots, 8, 9$ to the right end of an $(n-1)$-digit number without consecutive 1's, or we can add one of $\{01, 21, 31, 41, 51, 61, 71, 81, 91\}$ to the right end of an $(n-2)$-digit number with no consecutive 1's. This means that $w_n = 9w_{n-1} + 9w_{n-2}$.

Section 5.6 exercises

1. (a) The difference table below along with Theorem 2 can be used to conclude that $a_n = 7n + 3$.

n	0	1	2	3	4	5
a_n	3	10	17	24	31	$\cdots$
Δ_n^1	7	7	7	7	$\cdots$	$\cdots$

   (c) The difference table below along with Theorem 2 can be used to conclude that $a_n = 1 + 2n + \frac{3}{2}n(n-1)$.

n	0	1	2	3	4	5
a_n	1	3	8	16	27	41
Δ_n^1	2	5	8	11	14	$\cdots$
Δ_n^2	3	3	3	3	$\cdots$	$\cdots$

3. (a) The difference table below along with Theorem 2 can be used to conclude that the sum is equal to $n + 2n(n-1)$.

n	0	1	2	3	4	5
Σ_1^n	0	1	6	15	28	45
Δ_n^1	1	5	9	13	17	$\cdots$
Δ_n^2	4	4	4	4	$\cdots$	$\cdots$

   (c) The difference table below along with Theorem 2 can be used to conclude that the sum is equal to $6n + \frac{15}{2}n(n-1) + \frac{4}{3}n(n-1)(n-2)$.

n	0	1	2	3	4	5
Σ_1^n	0	6	27	71	146	260
Δ_n^1	6	21	44	75	114	$\cdots$
Δ_n^2	15	23	31	39	$\cdots$	$\cdots$
Δ_n^3	8	8	8	$\cdots$	$\cdots$	$\cdots$

5. $\Delta_0^0 = s_0$

6. (a) From Theorem 4, the solution must have the form $a_n = L \cdot 2^n - 3$. The initial condition $a_1 = 2$ tells us that $L = \frac{5}{2}$, and so $a_n = \frac{5}{2} \cdot 2^n - 3$.

(c) From Theorem 4, the solution must have the form $a_n = L \cdot (-2)^n + \frac{5}{3}$. The initial condition $a_1 = 0$ tells us that $L = \frac{5}{6}$, and so $a_n = \frac{5}{6}(-2)^n + \frac{5}{3}$

7. Let a_n = amount you owe after n months. Then $a_0 = 100,000$ and $a_n = (1.005)a_{n-1} - 1,000$, for $n \geq 1$, so $a_n = 100,000(2 - (1.005)^n)$

To find when the mortgage is paid, we solve $(1.005)^n = 2$. Since $\log_{1.005}(2) \approx 138.98$, we conclude that the mortgage will be paid off in the 139th month.

9. $s_0 = 5$ and $s_n = 3 \cdot s_{n-1} - 2$ for $n \geq 1$ has closed formula $s_n = \frac{4}{3} \cdot 3^n + 1$.

11. Since $a_n = L \cdot b^n + \frac{c}{1-b}$, if $a_0 = 1$, we must have $L = 1 - \frac{c}{1-b}$.

13. *Proof by induction.* If $n = 0$, the statement is "$s_0 = s_0 + K \cdot 0$," which is certainly true. Let $m \geq 1$ be given such that all of the statements up to "$s_{m-1} = s_0 + K \cdot (m-1)$" have been verified to be true. Now

$$
\begin{aligned}
s_m &= s_{m-1} + (s_m - s_{m-1}) \\
&= s_{m-1} + K \text{ since } s_n \text{ has a first different } K \\
&= (s_0 + K \cdot (m-1)) + K \\
&= s_0 + Km
\end{aligned}
$$

15. (c) The difference table looks like this:

n	0	1	2	3	4	5
a_n	1	3	10	29	74	173
Δ_n^1	2	7	19	45	99	$\cdots$
Δ_n^2	5	12	26	54	$\cdots$	$\cdots$
Δ_n^3	7	14	28	$\cdots$	$\cdots$	$\cdots$

Since $\Delta_n^3 = 7 \cdot 2^n$, we have

$$
\begin{aligned}
\Delta_n^3 &= 5 + \sum_{k=0}^{n-1} 7 \cdot 2^k \\
&= 5 + 7 \cdot (2^n - 1) = 7 \cdot 2^n - 2
\end{aligned}
$$

Since $\Delta_n^2 = 7 \cdot 2^n - 2$, we have

$$
\begin{aligned}
\Delta_n^1 &= 2 + \sum_{k=0}^{n-1} 7 \cdot 2^k - 2 \\
&= 2 + 7 \cdot (2^n - 1) - 2n \\
&= 7 \cdot 2^n - 2n - 5
\end{aligned}
$$

Therefore,

$$
\begin{aligned}
a_n &= 1 + \sum_{k=0}^{n-1} 7 \cdot 2^k - 2k - 5) \\
&= 1 + 7 \cdot (2^n - 1) - (n^2 - n) - 5n \\
&= 7 \cdot 2^n - n^2 - 4n - 6
\end{aligned}
$$

17. (a) $a_n = C \cdot 3^n + K \cdot 4^n$

(c) $a_n = C \cdot 4^n + K \cdot (-3)^n$

18. (a) $a_n = 4^n - 3^n$

    (c) $a_n = \frac{1}{7}(4^n - (-3)^n)$

19. (a) Since the equation $x^2 = x + 3$ has solutions $x = \frac{1 \pm \sqrt{13}}{2}$, it follows that the recurrence relation is satisfied by any sequence with closed form

$$a_n = C \cdot \left(\frac{1 + \sqrt{13}}{2}\right)^n + K \cdot \left(\frac{1 - \sqrt{13}}{2}\right)^n$$

20. (a) The conditions $a(0) = 1$ and $a(1) = 1$ lead to the equations $C + K = 1$ and $C \cdot \left(\frac{1+\sqrt{13}}{2}\right) + K \cdot \left(\frac{1-\sqrt{13}}{2}\right) = 1$ which have solutions $C = \frac{13+\sqrt{13}}{26}$ and $K = \frac{13-\sqrt{13}}{26}$. This leads to the closed formula for the recurrence relation

$$a_n = \left(\frac{13 + \sqrt{13}}{26}\right) \cdot \left(\frac{1 + \sqrt{13}}{2}\right)^n + \left(\frac{13 - \sqrt{13}}{26}\right) \cdot \left(\frac{1 - \sqrt{13}}{2}\right)^n$$

21. *Proof.* Suppose that the equation $x^2 = cx + d$ has distinct solutions $x = r$ and $x = s$, and consider the closed formula $a_n = C \cdot r^n + K \cdot s^n$. From this formula, it follows that $a_{n-1} = C \cdot r^{n-1} + K \cdot s^{n-1}$ and $a_{n-2} = C \cdot r^{n-2} + K \cdot s^{n-2}$. Now we can combine this information to get

$$
\begin{aligned}
c \cdot a_{n-1} + d \cdot a_{n-2} &= c \cdot (C \cdot r^{n-1} + K \cdot s^{n-1}) + d \cdot (C \cdot r^{n-2} + K \cdot s^{n-2}) \\
&= C \cdot (c \cdot r^{n-1} + d \cdot r^{n-2}) + K \cdot (c \cdot s^{n-1} + d \cdot s^{n-2}) \\
&= r^{n-2} \cdot C \cdot (c \cdot r + d) + s^{n-2} \cdot K \cdot (c \cdot s + d) \\
&= r^{n-2} \cdot (C \cdot r^2 + s^{n-2} \cdot K \cdot s^2) \\
&= C \cdot r^n + K \cdot s^n = a_n
\end{aligned}
$$

This verifies the relation $c \cdot a_{n-1} + d \cdot a_{n-2} = a_n$.

22. (a) Since the equation $x^2 = 4x - 4$ has unique solution $x = 2$, it follows that the recurrence relation is satisfied by any sequence with closed form

$$a_n = (C + K \cdot n) \cdot 2^n$$

23. (a) The conditions $a(0) = 1$ and $a(1) = 1$ lead to the equations $C \cdot 1 = 1$ and $(C + K) \cdot (2) = 1$ which have solutions $C = 1$ and $K = -\frac{1}{2}$. This leads to the closed formula for the recurrence relation

$$a_n = \left(1 - \frac{1}{2}n\right) \cdot 2^n$$

Section 6.1 exercises

1. (a) $\frac{13}{52} = \frac{1}{4}$

   (c) $\frac{8}{52} = \frac{2}{13}$

2. (a) $\frac{20}{100} = \frac{1}{5}$

3. (a) $\frac{10}{36} = \frac{5}{18}$

   (c) $\frac{6}{36} = \frac{1}{6}$

   (e) $\frac{36-6}{36} = \frac{30}{36} = \frac{5}{6}$

4. (a) $\frac{1}{2}$

5. Some have more than one answer.

   (a) The set of ordered lists of length 3 using elements from $\{1, 2, 3, 4, 5, 6\}$

   (b) The set of permutations of length 2 with elements from the set of club members (Treat the first person listed as president, the second as vice president.)

   (c) The set of ordered lists of length 6 with elements from $\{H, T\}$

   (d) The set of sets (combinations) of size 3 with elements from the set of 52 cards, OR the set of permutations of length 3 with elements from the set of 52 cards

7. (a) $\frac{4 \cdot 51}{52 \cdot 51} = \frac{1}{13}$

   (c) $\frac{1 \cdot 12 + 3 \cdot 13}{52 \cdot 51} = \frac{1}{52}$

   (e) $\frac{16 \cdot 51}{52 \cdot 51} = \frac{4}{13}$

9. There are a total of $C(23, 3) = 1771$ different committees possible.

   (a) $\frac{C(21,3)}{C(23,3)} = \frac{190}{253}$

   (c) $\frac{C(12,3)}{C(23,3)} = \frac{20}{161}$

11. $\frac{2 \cdot 4!}{5!} = \frac{2}{5}$

12. Our sample space will be the set of all combinations of size 5 with elements from the deck of 52 cards.

   (a) A flush can be constructed by (i) choosing a suit and (ii) choosing a set of five values from the 13 values available. This leads to the probability $\frac{4 \cdot C(13,5)}{C(52,5)} = \frac{33}{16\,660}$.

   (c) A straight flush can be constructed by (i) choosing a low card for the straight (which determines all 5 values for the straight) and (ii) choosing a single suit for the hand. This leads to the probability $\frac{10 \cdot 4}{C(52,5)} = \frac{1}{64\,974}$

15. $\frac{365^{32} - P(365, 32)}{365^{32}} \approx 75.3\%$

17. The probability that two people in a group of 500 have the same last digits of their Social Security Number is approximately 0.9999969.

19. (a) We can consider all outcomes of the two dice (as in Example 1) and count the number in which John's roll is higher. The probability is $\frac{15}{36} = \frac{5}{12}$.

   (c) We are assuming that John rolls an m-sided die and Jessica an n-sided die with $m \leq n$. Generalizing part (b), John wins in $1 + 2 + 3 + \cdots + (m-1) = \frac{m(m-1)}{2}$ of the $m \cdot n$ possible outcomes, so the probability that John wins is $\frac{m(m-1)}{2mn} = \frac{m-1}{2n}$.

21. $\frac{C(6,4)}{C(11,4)} \approx 0.045$

23. $\frac{C(n-1,31)}{C(n+31,31)}$

25. Answers will vary but should support the theoretical probability of $\frac{1}{3}$.

26. Answers will vary but should support the theoretical probability of approximately 0.22.

27. To see why this works, imagine one half (consisting of 13 red cards and 13 black cards) came from a blue-backed deck while the other half (consisting of 13 red cards and 13 black cards) came from a green-backed deck. The final deck will alternate colors of their backs even though the faces of the cards are fairly shuffled. As the solitaire game is played, cards are removed from the deck in adjacent pairs which share the same face-color. Hence at any point in the process, (i) the number of blue-backed red cards is equal to the number of green-backed red cards, (ii) the number of blue-backed black cards is equal to the number of green-backed black cards, and (iii) the blue-backed and green-backed cards alternate. Given these three properties that remain invariant as the game is played, it is impossible that the game should ever end in a loss. This is true because a losing final position would consist of cards alternating in face colors, and property (iii) then dictates that the red cards and black cards should have different back colors contrary to properties (i) and (ii).

Section 6.2 exercises

1. (a) A draw which includes the Ace of Clubs is in both events, so these events are not disjoint.

   (c) These events are disjoint.

2. (a) Let E_1 be the set of outcomes where the card is an Ace, and E_2 be the set of outcomes where the card is a Jack. Then since these events are disjoint,

$$Prob(E_1 \text{ or } E_2) = Prob(E_1) + Prob(E_2) = \frac{1}{13} + \frac{1}{13} = \frac{2}{13}$$

3. (a) Let E_1 be the set of outcomes where the card is an ace, and E_2 be the set of outcomes where the card is an heart. Then

$$\begin{aligned} Prob(E_1 \text{ or } E_2) &= Prob(E_1) + Prob(E_2) - Prob(E_1 \text{ and } E_2) \\ &= \frac{1}{13} + \frac{1}{4} - \frac{1}{52} = \frac{4}{13} \end{aligned}$$

   (c) Let E_1 be the set of outcomes where the card is a diamond or a club, and E_2 be the set of outcomes where the card is a king. Then

$$\begin{aligned} Prob(E_1 \text{ or } E_2) &= Prob(E_1) + Prob(E_2) - Prob(E_1 \text{ and } E_2) \\ &= \frac{1}{2} + \frac{1}{13} - \frac{1}{26} = \frac{4}{13} \end{aligned}$$

4. (a) These events are not independent. Let E_1 be the set of outcomes where the first card is an Ace, and E_2 be the set of outcomes where the second card is a Ten, Jack, Queen or King. Then $Prob(E_1) = \frac{1}{13}$, $Prob(E_2) = \frac{4}{13}$, but

$$Prob(E_1 \text{ and } E_2) = \frac{(4)(16)}{(52)(51)} = \frac{16}{663} \neq \frac{1}{13} \cdot \frac{4}{13}$$

   (c) These events are independent.

5. Because the first card is replaced and the deck is shuffled before the second card is drawn, in each case, the events described are independent.

   (a) $\frac{1}{13} \times \frac{1}{2} = \frac{1}{26}$

   (c) $\frac{9}{13} \times \frac{9}{13} = \frac{81}{169}$

7. $\frac{11}{36} + \frac{5}{36} - \frac{2}{36} = \frac{7}{18}$

9. Of the 6^3 possible outcomes, we can break the successful ones up by the result of the first die:

- **Case 1:** For the roll to look like $1,_-,_-$, the two blanks would have a sum between 4 and 9. There are $3 + 4 + 5 + 6 + 5 + 4 = 27$ such rolls.

- **Case 2:** For the roll to look like $2,_-,_-$, the two blanks would have a sum between 3 and 8. There are $2 + 3 + 4 + 5 + 6 + 5 = 25$ such rolls.

- **Case 3:** For the roll to look like $3,_-,_-$, the two blanks would have a sum between 2 and 7. There are $1 + 2 + 3 + 4 + 5 + 6 = 21$ such rolls.

- **Case 4:** For the roll to look like $4,_-,_-$, the two blanks would have a sum between 2 and 6. There are $1 + 2 + 3 + 4 + 5 = 15$ such rolls.

- **Case 5:** For the roll to look like $5,_-,_-$, the two blanks would have a sum between 2 and 5. There are $1 + 2 + 3 + 4 = 10$ such rolls.

- **Case 6:** For the roll to look like $6,_-,_-$, the two blanks would have a sum between 2 and 4. There are $1 + 2 + 3 = 6$ such rolls.

The probability then is $\frac{27}{6^3} + \frac{25}{6^3} + \frac{21}{6^3} + \frac{15}{6^3} + \frac{10}{6^3} + \frac{6}{6^3} = \frac{13}{27} \approx 0.48$

11. $\frac{1}{8} + \frac{1}{8} - \frac{1}{8} \cdot \frac{1}{8} = \frac{15}{64}$

13. $Prob(\text{sum is even}) = \frac{1}{2}$; $Prob(\text{sum is a multiple of 3}) = \frac{2+5+4+1}{36} = \frac{1}{3}$

15. $\frac{6!}{6^6} = \frac{5}{324}$

17. This uses the answers to Exercise 12 from Section 6.1 along with the general sum rule. Since $Prob(\text{flush}) = \frac{33}{16\,660}$, $Prob(\text{straight}) = \frac{128}{32\,487}$, and $Prob(\text{straight and flush}) = \frac{1}{64\,974}$, we have that the probability of a straight or a flush is $\frac{33}{16\,660} + \frac{128}{32\,487} - \frac{1}{64\,974} = \frac{1\,279}{216\,580}$.

19. (a) Let E_1 be the set of outcomes where the faceup card is an Ace, and let E_2 be the set of outcomes where the facedown card is a Ten, Jack, Queen or King. Then

$$Prob(E_2|E_1) = \frac{16}{51} \approx 0.314$$

(c) Let E_1 be the set of outcomes where the opponent's faceup card is an Ace and the next two cards (mine) are a Four and a Five, in either order, and let E_2 be the set of outcomes where the opponent's facedown card is a Ten, Jack, Queen or King. Then

$$Prob(E_2|E_1) = \frac{16}{49} \approx 0.327$$

21. (a) Let E_1 be the set of outcomes where the opponent's faceup cards are $\{2D,6C\}$, and let E_2 be the set of outcomes where the opponent's facedown cards have values $\{3,4,5\}$. Then

$$Prob(E_2|E_1) = \frac{C(12,3)}{C(50,3)} \approx 0.0112$$

(c) Let E_1 be the set of outcomes where the opponent's faceup cards are $\{5D,6C\}$, and let E_2 be the set of outcomes where the opponent's facedown cards have values $\{2,3,4\}$ or $\{3,4,7\}$ or $\{4,7,8\}$ or $\{7,8,9\}$. Then

$$Prob(E_2|E_1) = \frac{4 \cdot C(12,3)}{C(50,3)} \approx 0.0449$$

23. We use the numbers already calculated in Example 9.

(a) $Prob(P \text{ and } \overline{S})$ is the probability that the steroid result is positive and the athlete has not used steroids.

$$
\begin{aligned}
Prob(P \text{ and } \overline{S}) &= Prob(\overline{S}) \cdot Prob(P|\overline{S}) \\
&= (1 - Prob(S)) \cdot Prob(P|\overline{S}) \\
&= (0.03) \cdot (0.02) = 0.0006
\end{aligned}
$$

(b) Events $P \cap \overline{S}$ and $P \cap S$ are disjoint and $(P \cap \overline{S}) \cup (P \cap S) = P$, so by the sum rule,

$$Prob(P) = Prob(P \text{ and } S) + Prob(P \text{ and } \overline{S}) \approx 0.02985 + 0.0006 = 0.03045$$

(c) $Prob(S|P)$ is the probability that an athlete has used steroids given that the result of the test was positive.

$$Prob(S|P) = \frac{Prob(S \text{ and } P)}{Prob(P)} \approx \frac{0.02985}{0.03045} \approx 0.9803$$

83

(d) $Prob(S|\overline{P})$ is the probability that an athlete has used steroids given that the result of the test was negative.

$$
\begin{aligned}
Prob(S|\overline{P}) &= \frac{Prob(S \text{ and } \overline{P})}{Prob(\overline{P})} \\
&= \frac{Prob(\overline{P}|S) \cdot Prob(S)}{1 - Prob(P)} \\
&\approx \frac{(0.005)(0.03)}{1 - 0.03045} \approx 0.000155
\end{aligned}
$$

24. $C(4,2)(1/3)^2(2/3)^2 = \frac{8}{27} \approx 0.296$

26. $(C(6,0) + C(6,2) + C(6,4) + C(6,6))/2^6 = \frac{1}{2};\ C(6,0)(1/4)^6 + C(6,2)(3/4)^2(1/4)^4 + C(6,4)(3/4)^4(1/4)^2 + C(6,6)(3/4)^6 = \frac{65}{128} \approx 0.5078$

Section 6.3 exercises

1. $C(7,5) \cdot (1/6)^5 (5/6)^2 = \frac{175}{93\,312} \approx 0.001875$

3. $C(10,3) \cdot (1/2)^{10} = \frac{15}{128}$

5. $C(10,1) \cdot (1/6)^1 (5/6)^9 \approx 0.323$

7. $C(5,2)(1/6)^2(5/6)^3 + C(5,3)(1/6)^3(5/6)^2 + C(5,4)(1/6)^4(5/6)^1 + C(5,5)(1/6)^5$

9. There are several cases to consider. We use X to denote a roll of a die that is neither a 5 nor a 6.

Case 1. **One six and no fives.** These look like a 6 and three X's in any order, so the probability of this case happening is $C(4,1) \left(\frac{1}{6}\right) \left(\frac{4}{6}\right)^3$.

Case 2. **Two sixes and no fives.** These look like two 6's and two X's in any order, so the probability of this case happening is $C(4,2) \left(\frac{1}{6}\right)^2 \left(\frac{4}{6}\right)^2$.

Case 3. **Two sixes and one five.** These look like two 6's, one 5 and one X in any order, so the probability of this case happening is $C(4,2)C(2,1) \left(\frac{1}{6}\right)^3 \left(\frac{4}{6}\right)$.

Case 4. **Three sixes and no fives.** These look like three 6's and one X in any order, so the probability of this case happening is $C(4,3) \left(\frac{1}{6}\right)^3 \left(\frac{4}{6}\right)$.

Case 5. **Three sixes and one five.** These look like three 6's and one 5 in any order, so the probability of this case happening is $C(4,3) \left(\frac{1}{6}\right)^4$.

Case 6. **Four sixes.** The probability of this case happening is $\left(\frac{1}{6}\right)^4$.

The final probability that any of these cases occur is the sum of the values given in the six cases above:

$$
4\left(\frac{1}{6}\right)\left(\frac{4}{6}\right)^3 + 6\left(\frac{1}{6}\right)^2\left(\frac{4}{6}\right)^2 + 12\left(\frac{1}{6}\right)^3\left(\frac{4}{6}\right) + 4\left(\frac{1}{6}\right)^3\left(\frac{4}{6}\right) + 4\left(\frac{1}{6}\right)^4 + \left(\frac{1}{6}\right)^4
$$

This is approximately 0.325.

11. There are $C(52,5)$ hands in the sample space. We can create a "pair" by (i) choosing a pair value, (ii) choosing two suits for the pair, (iii) choosing three *other* values to complete the hand, and (iv) choosing the suit for each of these values from lowest to highest. Hence, $13 \cdot C(4,2) \cdot C(12,3) \cdot 4^3$ of the hands are a "pair," so the probability of getting a pair is

$$
\frac{13 \cdot C(4,2) \cdot C(12,3) \cdot 4^3}{C(52,5)} = \frac{352}{833} \approx 0.4226
$$

14. Let p be the probability that the Bears win an individual game. Then the probability that the Bears win a best-of-three series is

$$p^2 + 2p^2(1-p)$$

Using technology we can show that this is at least 0.9 when $p \geq 0.8042$.

16. Here's a graph created with a spreadsheet:

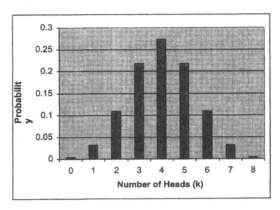

18. $C(9,3) \cdot C(6,3) \cdot C(3,3) \cdot (1/3)^3 \cdot (1/2)^3 \cdot (1/6)^3 = \frac{35}{972} \approx 0.036$

19. (a) To get at least two hits means two hits or three hits or four hits or five hits, so we add

$$C(5,2)(1/3)^2(2/3)^3 + C(5,3)(1/3)^3(2/3)^2 + C(5,4)(1/3)^4(2/3)^1 + C(5,5)(1/3)^5 = \frac{131}{243} \approx 0.539.$$

(c) We represent five plate appearances as an ordered list of length five using elements from $\{h,b,n\}$, where h represents a hit, b represents a base-on-balls, and n represents anything else. We treat each of the nine possible cases where the given condition is satisfied.

- 1 h, 0 b's, 4 n's $C(5,1)$ of these, each occurring with probability $(1/3)^1(1/6)^0(1/2)^4$.
- 3 h's, 0 b's, 2 n's $C(5,3)$ of these, each occurring with probability $(1/3)^3(1/6)^0(1/2)^2$.
- 4 h's, 0 b's, 1 n $C(5,4)$ of these, each occurring with probability $(1/3)^4(1/6)^0(1/2)^1$.
- 2 h's, 0 b's, 3 n's $C(5,2)$, each occurring with probability $(1/3)^2(1/6)^0(1/2)^3$.
- 3 h's, 1 b, 1 n $C(5,3)C(2,1)$, each occurring with probability $(1/3)^3(1/6)^1(1/2)^1$.
- 4 h's, 1 b, 0 n's $C(5,4)$, each occurring with probability $(1/3)^4(1/6)^1(1/2)^0$.
- 2 h's, 1 b, 2 n's $C(5,2)C(3,1)$, each occurring with probability $(1/3)^2(1/6)^1(1/2)^2$.
- 3 h's, 2 b's, 0 n's $C(5,3)$, each occurring with probability $(1/3)^3(1/6)^2(1/2)^0$.
- 5 h's, 0 b's, 0 n's $C(5,5)$, occurring with probability $(1/3)^5(1/6)^0(1/2)^0$.

So the answer is the sum

$$
\begin{aligned}
&C(5,1)(1/3)^1(1/6)^0(1/2)^4 + C(5,3)(1/3)^3(1/6)^0(1/2)^2 + \\
&C(5,4)(1/3)^4(1/6)^0(1/2)^1 + C(5,2)(1/3)^2(1/6)^0(1/2)^3 + \\
&C(5,3)C(2,1)(1/3)^3(1/6)^1(1/2)^1 + C(5,4)(1/3)^4(1/6)^1(1/2)^0 + \\
&C(5,2)C(3,1)(1/3)^2(1/6)^1(1/2)^2 + C(5,3)(1/3)^3(1/6)^2(1/2)^0 + \\
&C(5,5)(1/3)^5(1/6)^0(1/2)^0 \\
&= \frac{767}{1296} \approx 0.59.
\end{aligned}
$$

22. (a) $(3/5)^2 + C(2,1) \cdot (3/5)^2(2/5) = \frac{81}{125} \approx 0.648$

(b) $(3/5)^3 + C(3,1) \cdot (3/5)^3(2/5) + C(4,2) \cdot (3/5)^3(2/5)^2 = \frac{2133}{3125} \approx 0.683$

(c) $(3/5)^4 + C(4,1) \cdot (3/5)^4(2/5) + C(5,2) \cdot (3/5)^4(2/5)^2 + C(6,3) \cdot (3/5)^4(2/5)^3 = \frac{11097}{15625} \approx 0.710$

23. There are many different outcomes. We list them all and find the probability of each.

Outcome	Probability	Outcome	Probability	Outcome	Probability
AAAA	$(3/5)^2(2/5)^2$	ABBAAA	$(3/5)^3(2/5)^3$	BBAAABA	$(2/5)^6(3/5)$
AAABA	$(3/5)^3(2/5)^2$	BABAAA	$(3/5)^3(2/5)^3$	AABBBAA	$(3/5)^7$
AABAA	$(3/5)^3(2/5)^2$	BBAAAA	$(2/5)^5(3/5)$	ABABBAA	$(3/5)^5(2/5)^2$
ABAAA	$(3/5)(2/5)^4$	AAABBBA	$(3/5)^5(2/5)^2$	BAABBAA	$(3/5)^5(2/5)^2$
BAAAA	$(3/5)(2/5)^4$	AABABBA	$(3/5)^5(2/5)^2$	ABBABAA	$(3/5)^5(2/5)^2$
AAABBA	$(3/5)^5(2/5)$	ABAABBA	$(3/5)^3(2/5)^4$	BABABAA	$(3/5)^5(2/5)^2$
AABABA	$(3/5)^5(2/5)$	BAAABBA	$(3/5)^3(2/5)^4$	ABBBAAA	$(3/5)^5(2/5)^2$
ABAABA	$(3/5)^3(2/5)^3$	AABBABA	$(3/5)^5(2/5)^2$	BABBAAA	$(3/5)^5(2/5)^2$
BAAABA	$(3/5)^3(2/5)^3$	ABABABA	$(3/5)^3(2/5)^4$	BBABAAA	$(3/5)^3(2/5)^4$
AABBAA	$(3/5)^5(2/5)$	BAABABA	$(3/5)^3(2/5)^4$	BBAABAA	$(3/5)^3(2/5)^4$
ABABAA	$(3/5)^3(2/5)^3$	ABBAABA	$(3/5)^3(2/5)^4$	BBBAAAA	$(3/5)^3(2/5)^4$
BAABAA	$(3/5)^3(2/5)^3$	BABAABA	$(3/5)^3(2/5)^4$		

The sum of these is $(3/5)^2(2/5)^2+2(3/5)^3(2/5)^2+2(3/5)(2/5)^4+3(3/5)^5(2/5)+6(3/5)^3(2/5)^3+(3/5)(2/5)^5+$
$9(3/5)^5(2/5)^2 + 9(3/5)^3(2/5)^4 + (3/5)(2/5)^5 + (3/5)^7 = \frac{41853}{78125} \approx 0.536$.

Section 6.4 exercises

1. Letting X denote the sum of a pair of dice, we have in the case of fair dice

$$E[X] = \frac{1}{36}(2) + \frac{2}{36}(3) + \frac{3}{36}(4) + \frac{4}{36}(5) + \frac{5}{36}(6) + \frac{6}{36}(7)$$
$$+ \frac{5}{36}(8) + \frac{4}{36}(9) + \frac{3}{36}(10) + \frac{2}{36}(11) + \frac{1}{36}(12)$$
$$= 7$$

If we use the loaded dice that come up "6" with probability $\frac{1}{2}$ and every other face with probability $\frac{1}{10}$, we have

$$E[X] = \frac{1}{100}(2) + \frac{2}{100}(3) + \frac{3}{100}(4) + \frac{4}{100}(5) + \frac{5}{100}(6) + \frac{14}{100}(7)$$
$$+ \frac{13}{100}(8) + \frac{12}{100}(9) + \frac{11}{100}(10) + \frac{10}{100}(11) + \frac{25}{100}(12)$$
$$= 9$$

In either case, we get twice the expected value of a single die roll. This supports the intuition for the basic rule $E[A + B] = E[A] + E[B]$.

3. $0 \cdot \frac{C(30,3)}{C(50,3)} + 1 \cdot \frac{C(20,1)C(30,2)}{C(50,3)} + 2 \cdot \frac{C(20,2)C(30,1)}{C(50,3)} + 3 \cdot \frac{C(20,3)}{C(50,3)} = 1.2$

5. $(1) \cdot \frac{C(4,1) \cdot C(48,4)}{C(52,5)} + (2) \cdot \frac{C(4,2) \cdot C(48,3)}{C(52,5)} + (3) \cdot \frac{C(4,3) \cdot C(48,2)}{C(52,5)} + (4) \cdot \frac{C(4,4) \cdot C(48,1)}{C(52,5)} = \frac{5}{13} \approx 0.385$

6. Let X be the sum of the values of the pair of cards. X can be any number from $\{4, 5, 6, \ldots, 22\}$.

Sum	Probability	Sum	Probability	Sum	Probability	Sum	Probability
4	6/1326	9	48/1326	14	118/1376	19	80/1376
5	16/1326	10	54/1326	15	112/1376	20	136/1376
6	22/1326	11	64/1326	15	102/1376	21	64/1376
7	32/1326	12	118/1326	17	96/1376	22	6/1376
8	38/1326	13	128/1326	18	86/1376		

So
$$E[X] = (4)\left(\frac{6}{1326}\right) + (5)\left(\frac{16}{1326}\right) + (6)\left(\frac{22}{1326}\right) + \cdots + (22)\left(\frac{6}{1326}\right) \approx 14.88$$

For those interested in card games, this is the value of the average starting hand in Blackjack (counting Ace as 11).

9. The expected payoff is $10 \cdot \left(\frac{75}{216}\right) + 25 \cdot \left(\frac{15}{216}\right) + 45 \cdot \left(\frac{1}{216}\right) \approx \$5.42$. If 10 000 people play, the casino will take in \$100 000 and expect to pay out about \$54 200, so the casino expects to make a profit of about \$45 800 per day on this game.

11. Using the hint, we recall the identity from Exercise 36 of Section 5.3:

$$\sum_{k=1}^{6} k \cdot C(6, k) = 6 \cdot 2^5$$

Now using X to denote the number of 1's we see in our tossing of six coins, we have

$$\begin{aligned} E[X] &= (0)C(6,0)\left(\frac{1}{2}\right)^6 + (1)C(6,1)\left(\frac{1}{2}\right)^6 + \cdots + (6)C(6,6)\left(\frac{1}{2}\right)^6 \\ &= \left(\frac{1}{2}\right)^6 \left(\sum_{k=0}^{6} k \cdot C(6,k)\right) \\ &= \left(\frac{1}{2}\right)^6 (6 \cdot 2^5) = 3 \end{aligned}$$

13. Since the general identity from Exercise 37 of Section 5.3 is

$$\sum_{k=1}^{n} k \cdot C(n,k) \cdot x^k = n \cdot x \cdot (1+x)^{n-1}$$

we can use $x = 1/5$ and $n = 10$ to get

$$\sum_{k=1}^{10} k \cdot C(10,k) \cdot (1/5)^k = 10 \cdot (1/5) \cdot \left(\frac{6}{5}\right)^9$$

and we can multiply both sides of this equation by $(5/6)^{10}$ (and note that $\left(\frac{1}{5}\right)^k \left(\frac{5}{6}\right)^{10} = \left(\frac{1}{6}\right)^k \left(\frac{5}{6}\right)^{10-k}$) to get

$$\sum_{k=1}^{10} k \cdot C(10,k) \cdot \left(\frac{1}{6}\right)^k \left(\frac{5}{6}\right)^{10-k} = 2 \cdot \frac{5}{6}$$

Now if X is the number of "1"s that occur in the ten rolls, then

$$\begin{aligned} E[X] &= (1)C(10,1)\left(\frac{1}{6}\right)^1 \left(\frac{5}{6}\right)^9 + \cdots + (10)C(10,10)\left(\frac{1}{6}\right)^{10}\left(\frac{5}{6}\right)^0 \\ &= \sum_{k=1}^{10} k \cdot C(10,k) \cdot \left(\frac{1}{6}\right)^k \left(\frac{5}{6}\right)^{10-k} \\ &= 2 \cdot \frac{5}{6} = \frac{5}{3} \end{aligned}$$

by the identity above.

15. Knowing that at least one child is a boy means we have sample space $\{MF, FM, MM\}$, in which case the expected number of boys is $1 \cdot 2/3 + 2 \cdot 1/3 = 4/3$. Knowing that the older child is a boy means we have the sample space $\{MF, MM\}$, in which case the expected number of boys is $1 \cdot 1/2 + 2 \cdot 1/2 = 3/2$. Is it

surprising that these are not equal? Imagine visiting 6 000 homes in which you know that the family has two children and at least one is a boy. Of these, it is reasonable to expect that about 2 000 homes have two boys, about 2 000 homes have an older boy and a younger girl, and about 2 000 homes have a younger boy and an older girl. The average number of boys per home would be about $\frac{4\,000+2\,000+2\,000}{6\,000} \approx \frac{4}{3}$. If we imagine visiting 6 000 other homes in which we know ahead of time there are two children, the older of which is a boy, we expect the average number of boys per home to be about $\frac{6\,000+3\,000}{6\,000} = \frac{3}{2}$. Note that the higher probability of having two boys is the main reason that the answer in the latter case is larger.

17. We can break the problem down just as in Exercise 16 to get

$$3((2/3)^3 + (1/3)^3) + 4(3(2/3)^3(1/3) + 3(1/3)^3(2/3)) + 5(6(2/3)^3(1/3)^2 + 6(1/3)^3(2/3)^2) = \frac{107}{27} \approx 3.96$$

19. Following the method of Example 7 with X representing the number of games in the best-of-three series, we have

- $Prob(X = 3) = p^3 + (1-p)^3$
- $Prob(X = 4) = 3p^3(1-p) + 3p(1-p)^3$
- $Prob(X = 5) = 6p^3(1-p)^2 + 6p^2(1-p)^3$

and so

$$
\begin{aligned}
E[X] &= 3 \cdot \left(p^3 + (1-p)^3\right) + 4 \cdot \left(3p^3(1-p) + 3p(1-p)^3\right) + 5 \cdot \left(6p^3(1-p)^2 + 6p^2(1-p)^3\right) \\
&= 6p^4 - 12p^3 + 3p^2 + 3p + 3
\end{aligned}
$$

The graph of this expression shows that the maximum length of the series occurs when $p = 1/2$.

21. The solution to Example 7 tells us that the average length of a best-of-seven series between two equally matched opponents is 5.8125. Since the winner of the series must win 4 games, this means that the losing team wins an average of 1.8125 games. This constitutes an average winning margin of 2.1875 games.

23. If we "zoom in" on the graph in the solution to Example 7, we can see that when p is less than about 0.215 or more than about 0.785, the average length of a best-of-seven series is less than 5 games.

Section 6.5 exercises

1. Letting X denote the number of rolls until a "1" is rolled for the first time, we have

$$E[X] = \frac{1}{6}(1) + \frac{5}{6}(1 + E[X])$$

Solving this equation algebraically gives us $E[X] = 6$.

3. Letting X denote the number of rolls until at least one "1" occurs for the first time, we have

$$E[X] = \frac{11}{36}(1) + \frac{25}{36}(1 + E[X])$$

Solving this equation algebraically gives us $E[X] = \frac{36}{11}$.

5. (a) Use p_n to denote the probability that the nth card is the first "Ace." Then for example, $p_1 = \frac{1}{13}$, $p_2 = \frac{48}{52} \cdot \frac{1}{13}$, $p_3 = \frac{48}{52} \cdot \frac{47}{51} \cdot \frac{1}{13}$, and in general,

$$p_n = \frac{P(48, n-1)}{P(52, n-1)} \cdot \frac{1}{13}$$

(c) The solution to Exercise 4 assumes that the trials are independent. "Drawing cards without replacement" does not have this property.

7. We have the recurrence relation $h_1 = 2$ and

$$h_n = \left(\frac{1}{2}\right)(1 + h_{n-1}) + \left(\frac{1}{2}\right)(1 + h_n)$$

or more simply, $h_n = 2 + h_{n-1}$, for all $n \geq 2$.

9. Let p denote the probability of the server eventually winning a game currently tied at deuce. Then

$$p = \left(\frac{9}{10}\right)^2 + 2 \cdot \left(\frac{9}{10}\right)\left(\frac{1}{10}\right) \cdot p$$

which has solution $p = 81/82$.

11. The expected length satisfies the equation

$$E[X] = 2 \cdot \left(\frac{3}{4}\right)^2 + 2 \cdot \left(\frac{1}{4}\right)^2 + 2 \cdot \left(\frac{3}{4}\right)\left(\frac{1}{4}\right)(2 + E[X])$$

which has solution $E[X] = \frac{16}{5} = 3.2$.

13. Let Q denote the probability of the server eventually winning a game currently tied at deuce. Then

$$Q = p^2 + 2 \cdot p \cdot (1 - p) \cdot Q$$

which has solution $Q = \frac{p^2}{p^2 + (1-p)^2}$. This is consistent with what we have seen in the previous exercises. On the other hand, the expected length satisfies the equation

$$E[X] = 2 \cdot p^2 + 2 \cdot (1 - p)^2 + 2 \cdot p(1 - p)(2 + E[X])$$

which has solution $E[X] = \frac{2}{p^2 + (1-p)^2}$. This is also consistent with what we have seen in the previous exercises.

15. We first consider games that are never tied at deuce.

- $Prob(A \text{ wins } 4 - 0) = C(3,0) \cdot \left(\frac{2}{3}\right)^4 = \frac{16}{81}$.
- $Prob(A \text{ wins } 4 - 1) = C(4,1) \cdot \left(\frac{2}{3}\right)^4 \left(\frac{1}{3}\right) = \frac{64}{243}$.
- $Prob(A \text{ wins } 4 - 2) = C(5,2) \cdot \left(\frac{2}{3}\right)^4 \left(\frac{1}{3}\right)^2 = \frac{160}{729}$.

If a game is tied at deuce, then the probability p that A eventually wins satisfies the equation

$$p = \left(\frac{2}{3}\right)^2 + 2 \cdot \left(\frac{2}{3}\right) \cdot \left(\frac{1}{3}\right) \cdot p$$

which has solution $p = 4/5$. The probability that the game first reaches deuce (at $3 - 3$) is $C(6,3) \cdot \left(\frac{2}{3}\right)^3 \left(\frac{1}{3}\right)^3 = \frac{160}{729}$. Therefore, the probability that A wins the game is

$$\frac{16}{81} + \frac{64}{243} + \frac{160}{729} + \left(\frac{160}{729} \cdot \frac{4}{5}\right) = \frac{208}{243} \approx 0.856$$

17. (c) Solving these three equations simultaneously gives us $a = p = 5$ and $t = 9$.

19. Using the formula derived in Example 3, the expected length of this game is $6 \cdot 8 = 48$ coin tosses.

21. In the Hank and Ted game played with a fair coin, if Hank starts with X markers and Ted starts with Y markers, then the probability that Hank wins is $\frac{X}{X+Y}$, the probability that Ted wins is $\frac{Y}{X+Y}$, and the expected length of the game is $\underline{(X)(Y)}$ moves.

23. Let $d_n = (n)(M - n)$. The conditions $d_0 = 0$ and $d_M = 0$ are obviously satisfied by this definition. Since

$$d_{n+1} + 1 = (n + 1)(M - (n + 1)) + 1 = nM - n^2 - 2n + M$$

and

$$d_{n-1} + 1 = (n - 1)(M - (n - 1)) + 1 = nM - n^2 + 2n + M$$

we conclude that

$$
\begin{aligned}
\frac{1}{2}(d_{n+1} + 1) + \frac{1}{2}(d_{n-1} + 1) &= \frac{1}{2}((nM - n^2 - 2n + M) + (nM - n^2 + 2n + M)) \\
&= \frac{1}{2}(2nM - 2n^2 + 2M) \\
&= nM - n^2 + M + 1 = n(M - n) = d_n
\end{aligned}
$$

which is Equation (6.2).

25. (a) Let $d_n = c\left(1 - \frac{1}{2^n}\right) - 3n$. Then $\frac{2}{3}(d_{n+1} + 1) + \frac{1}{3}(d_{n-1} + 1)$ can be simplifed as

$$
\begin{aligned}
&\frac{2}{3}\left(c\left(1 - \frac{1}{2^{n+1}}\right) - 3(n + 1) + 1\right) + \frac{1}{3}\left(c\left(1 - \frac{1}{2^{n-1}}\right) - 3(n - 1) + 1\right) \\
&= \left(\frac{2c}{3} - \frac{1}{3}\left(\frac{c}{2^n}\right) - 2n - \frac{4}{3}\right) + \left(\frac{c}{3} - \frac{2}{3}\left(\frac{c}{2^n}\right) - n + \frac{4}{3}\right) \\
&= c - \frac{c}{2^n} - 3n = c\left(1 - \frac{1}{2^n}\right) - 3n
\end{aligned}
$$

This final simplification is precisely the definition of d_n, so the relationship holds.

(b) For $d_n = c\left(1 - \frac{1}{2^n}\right) - 3n$, the equation $d_{20} = 0$ is the same as

$$c\left(\frac{1}{2}\right)^{20} - 3(20) = 0$$

which has solution $c \approx 60.00005722$.

(c) Using the previous result,

$$d_{10} = 60.00005722 \cdot \left(1 - \frac{1}{2^{10}}\right) - 3(10) \approx 29.94$$

(d) Looking at all values of d_n for $0 \le n \le 20$, we see that the largest is $d_4 \approx 44.25$. That is, the game that begins with $H = 4, T = 16$ is expected to last the longest.

26. This is a trick question. Generalizing the solution to Exercise 24, in the game with M total markers, the probability of Hank winning the game in which he starts with 10 markers is

$$p_{10} = \frac{2^{M-1}}{2^M - 1} \cdot \frac{2^{10} - 1}{2^9}$$

When M is large, the value of $\frac{2^{M-1}}{2^M-1}$ is very slightly larger than 0.5, and of course, $\frac{2^{10}-1}{2^9}$ is slightly larger than 1.998. Hence for large values of M, p_{10} is slightly larger than 0.999. That is, no matter how many total markers Ted has, Hank will win this game with probability greater than 0.999.

Section 6.6 exercises

1. For each i with $1 \le i \le 7$, State i will be the game where Hank has $i - 1$ markers and Ted has $7 - i$ markers.

$$
M = \begin{bmatrix}
1 & 0 & 0 & 0 & 0 & 0 & 0 \\
2/3 & 0 & 1/3 & 0 & 0 & 0 & 0 \\
0 & 2/3 & 0 & 1/3 & 0 & 0 & 0 \\
0 & 0 & 2/3 & 0 & 1/3 & 0 & 0 \\
0 & 0 & 0 & 2/3 & 0 & 1/3 & 0 \\
0 & 0 & 0 & 0 & 2/3 & 0 & 1/3 \\
0 & 0 & 0 & 0 & 0 & 0 & 1
\end{bmatrix}
$$

3. Define states as follows:

State 1	Andre wins	State 4	Andre up 1	State 7	Pete up 2
State 2	Andre up 3	State 5	Tied	State 8	Pete up 3
State 3	Andre up 2	State 6	Pete up 1	State 9	Pete wins

$$M = \begin{bmatrix} 1 & 0 & 0 & 0 & 0 & 0 & 0 & 0 & 0 \\ 2/3 & 0 & 1/3 & 0 & 0 & 0 & 0 & 0 & 0 \\ 0 & 2/3 & 0 & 1/3 & 0 & 0 & 0 & 0 & 0 \\ 0 & 0 & 2/3 & 0 & 1/3 & 0 & 0 & 0 & 0 \\ 0 & 0 & 0 & 2/3 & 0 & 1/3 & 0 & 0 & 0 \\ 0 & 0 & 0 & 0 & 2/3 & 0 & 1/3 & 0 & 0 \\ 0 & 0 & 0 & 0 & 0 & 2/3 & 0 & 1/3 & 0 \\ 0 & 0 & 0 & 0 & 0 & 0 & 2/3 & 0 & 1/3 \\ 0 & 0 & 0 & 0 & 0 & 0 & 0 & 0 & 1 \end{bmatrix}$$

5. States 1 through 6 will refer to the game piece being on squares A through F, respectively. The transition matrix is

$$M = \begin{bmatrix} 0 & 1/4 & 1/2 & 1/4 & 0 & 0 \\ 1/4 & 0 & 1/4 & 1/4 & 1/4 & 0 \\ 1/4 & 0 & 0 & 1/2 & 1/4 & 0 \\ 1/4 & 0 & 0 & 1/4 & 1/4 & 1/4 \\ 0 & 0 & 0 & 1/4 & 0 & 3/4 \\ 0 & 0 & 0 & 0 & 0 & 1 \end{bmatrix}$$

7. Answers vary but should support Exercises 12 and 24.

11. Answers vary but should support Exercise 28.

13. The probability of going from State 5 to State 1 in no more than 16 moves is approximately 0.0642, since this is the entry in Row 5, Column 1 of the matrix

$$M^{16} = \begin{bmatrix} 1.0 & 0.0 & 0.0 & 0.0 & 0.0 & 0.0 & 0.0 & 0.0 & 0.0 \\ 0.530 & 0.00741 & 0.0178 & 0.0306 & 0.0423 & 0.0510 & 0.0501 & 0.0359 & 0.235 \\ 0.273 & 0.00990 & 0.0244 & 0.0413 & 0.0590 & 0.0702 & 0.0710 & 0.0501 & 0.401 \\ 0.135 & 0.00945 & 0.0230 & 0.0402 & 0.0568 & 0.0701 & 0.0702 & 0.0510 & 0.544 \\ 0.0642 & 0.00726 & 0.0182 & 0.0316 & 0.0463 & 0.0568 & 0.0590 & 0.0423 & 0.674 \\ 0.0288 & 0.00486 & 0.0120 & 0.0216 & 0.0316 & 0.0402 & 0.0413 & 0.0306 & 0.789 \\ 0.0118 & 0.00265 & 0.00676 & 0.0120 & 0.0182 & 0.0230 & 0.0244 & 0.0178 & 0.883 \\ 0.00384 & 0.00106 & 0.00265 & 0.00486 & 0.00726 & 0.00945 & 0.00990 & 0.00741 & 0.954 \\ 0.0 & 0.0 & 0.0 & 0.0 & 0.0 & 0.0 & 0.0 & 0.0 & 1.0 \end{bmatrix}$$

15. Using the following definition of the fifteen states of this game, the transition matrix M is given below.

State 1:	$A = 0, B = 0$	State 6:	$A = 1, B = 1$	State 11:	$A = 2, B = 2$
State 2:	$A = 1, B = 0$	State 7:	$A = 2, B = 1$	State 12:	$A = 3, B = 2$
State 3:	$A = 2, B = 0$	State 8:	$A = 3, B = 1$	State 13:	$A = 0, B = 3$
State 4:	$A = 3, B = 0$	State 9:	$A = 0, B = 2$	State 14:	$A = 1, B = 3$
State 5:	$A = 0, B = 1$	State 10:	$A = 1, B = 2$	State 15:	$A = 2, B = 3$

$$M = \begin{bmatrix} 0 & \frac{1}{2} & 0 & 0 & \frac{1}{2} & 0 & 0 & 0 & 0 & 0 & 0 & 0 & 0 & 0 & 0 \\ 0 & 0 & \frac{1}{2} & 0 & 0 & \frac{1}{2} & 0 & 0 & 0 & 0 & 0 & 0 & 0 & 0 & 0 \\ 0 & 0 & 0 & \frac{1}{2} & 0 & 0 & \frac{1}{2} & 0 & 0 & 0 & 0 & 0 & 0 & 0 & 0 \\ 0 & 0 & 0 & 1 & 0 & 0 & 0 & 0 & 0 & 0 & 0 & 0 & 0 & 0 & 0 \\ 0 & 0 & 0 & 0 & 0 & \frac{1}{2} & 0 & 0 & \frac{1}{2} & 0 & 0 & 0 & 0 & 0 & 0 \\ 0 & 0 & 0 & 0 & 0 & 0 & \frac{1}{2} & 0 & 0 & \frac{1}{2} & 0 & 0 & 0 & 0 & 0 \\ 0 & 0 & 0 & 0 & 0 & 0 & 0 & \frac{1}{2} & 0 & 0 & \frac{1}{2} & 0 & 0 & 0 & 0 \\ 0 & 0 & 0 & 0 & 0 & 0 & 0 & 1 & 0 & 0 & 0 & 0 & 0 & 0 & 0 \\ 0 & 0 & 0 & 0 & 0 & 0 & 0 & 0 & 0 & \frac{1}{2} & 0 & 0 & \frac{1}{2} & 0 & 0 \\ 0 & 0 & 0 & 0 & 0 & 0 & 0 & 0 & 0 & 0 & \frac{1}{2} & 0 & 0 & \frac{1}{2} & 0 \\ 0 & 0 & 0 & 0 & 0 & 0 & 0 & 0 & 0 & 0 & 0 & \frac{1}{2} & 0 & 0 & \frac{1}{2} \\ 0 & 0 & 0 & 0 & 0 & 0 & 0 & 0 & 0 & 0 & 0 & 1 & 0 & 0 & 0 \\ 0 & 0 & 0 & 0 & 0 & 0 & 0 & 0 & 0 & 0 & 0 & 0 & 1 & 0 & 0 \\ 0 & 0 & 0 & 0 & 0 & 0 & 0 & 0 & 0 & 0 & 0 & 0 & 0 & 1 & 0 \\ 0 & 0 & 0 & 0 & 0 & 0 & 0 & 0 & 0 & 0 & 0 & 0 & 0 & 0 & 1 \end{bmatrix}$$

The matrix showing just the transient states (1, 2, 3, 5, 6, 7, 9, 10 and 11) is

$$N = \begin{bmatrix} 0 & \frac{1}{2} & 0 & \frac{1}{2} & 0 & 0 & 0 & 0 & 0 \\ 0 & 0 & \frac{1}{2} & 0 & \frac{1}{2} & 0 & 0 & 0 & 0 \\ 0 & 0 & 0 & 0 & 0 & \frac{1}{2} & 0 & 0 & 0 \\ 0 & 0 & 0 & 0 & \frac{1}{2} & 0 & \frac{1}{2} & 0 & 0 \\ 0 & 0 & 0 & 0 & 0 & \frac{1}{2} & 0 & \frac{1}{2} & 0 \\ 0 & 0 & 0 & 0 & 0 & 0 & 0 & 0 & \frac{1}{2} \\ 0 & 0 & 0 & 0 & 0 & 0 & 0 & \frac{1}{2} & 0 \\ 0 & 0 & 0 & 0 & 0 & 0 & 0 & 0 & \frac{1}{2} \\ 0 & 0 & 0 & 0 & 0 & 0 & 0 & 0 & 0 \end{bmatrix}$$

from which we can compute

$$N + N^2 + N^3 + N^4 + N^5 = \begin{bmatrix} 0 & 1/2 & 1/4 & 1/2 & 1/2 & 3/8 & 1/4 & 3/8 & 3/8 \\ 0 & 0 & 1/2 & 0 & 1/2 & 1/2 & 0 & 1/4 & 3/8 \\ 0 & 0 & 0 & 0 & 0 & 1/2 & 0 & 0 & 1/4 \\ 0 & 0 & 0 & 0 & 1/2 & 1/4 & 1/2 & 1/2 & 3/8 \\ 0 & 0 & 0 & 0 & 0 & 1/2 & 0 & 1/2 & 1/2 \\ 0 & 0 & 0 & 0 & 0 & 0 & 0 & 0 & 1/2 \\ 0 & 0 & 0 & 0 & 0 & 0 & 0 & 1/2 & 1/4 \\ 0 & 0 & 0 & 0 & 0 & 0 & 0 & 0 & 1/2 \\ 0 & 0 & 0 & 0 & 0 & 0 & 0 & 0 & 0 \end{bmatrix}$$

Adding 1 to the sum of the entries in the first row (corresponding to games beginning in State 1) gives us $\frac{33}{8} \approx 4.125$ games expected in this best-of-five series.

17. Using the definition of the states of the game given in the solution to Exercise 15, the transition matrix for this game is

$$M = \begin{bmatrix}
0 & \frac{2}{3} & 0 & 0 & \frac{1}{3} & 0 & 0 & 0 & 0 & 0 & 0 & 0 & 0 & 0 \\
0 & 0 & \frac{2}{3} & 0 & 0 & \frac{1}{3} & 0 & 0 & 0 & 0 & 0 & 0 & 0 & 0 \\
0 & 0 & 0 & \frac{2}{3} & 0 & 0 & \frac{1}{3} & 0 & 0 & 0 & 0 & 0 & 0 & 0 \\
0 & 0 & 0 & 1 & 0 & 0 & 0 & 0 & 0 & 0 & 0 & 0 & 0 & 0 \\
0 & 0 & 0 & 0 & 0 & \frac{2}{3} & 0 & 0 & \frac{1}{3} & 0 & 0 & 0 & 0 & 0 \\
0 & 0 & 0 & 0 & 0 & 0 & \frac{2}{3} & 0 & 0 & \frac{1}{3} & 0 & 0 & 0 & 0 \\
0 & 0 & 0 & 0 & 0 & 0 & 0 & \frac{2}{3} & 0 & 0 & \frac{1}{3} & 0 & 0 & 0 \\
0 & 0 & 0 & 0 & 0 & 0 & 0 & 1 & 0 & 0 & 0 & 0 & 0 & 0 \\
0 & 0 & 0 & 0 & 0 & 0 & 0 & 0 & 0 & \frac{2}{3} & 0 & 0 & \frac{1}{3} & 0 \\
0 & 0 & 0 & 0 & 0 & 0 & 0 & 0 & 0 & 0 & \frac{2}{3} & 0 & 0 & \frac{1}{3} \\
0 & 0 & 0 & 0 & 0 & 0 & 0 & 0 & 0 & 0 & 0 & \frac{2}{3} & 0 & 0 & \frac{1}{3} \\
0 & 0 & 0 & 0 & 0 & 0 & 0 & 0 & 0 & 0 & 0 & 1 & 0 & 0 & 0 \\
0 & 0 & 0 & 0 & 0 & 0 & 0 & 0 & 0 & 0 & 0 & 0 & 1 & 0 \\
0 & 0 & 0 & 0 & 0 & 0 & 0 & 0 & 0 & 0 & 0 & 0 & 0 & 1 & 0 \\
0 & 0 & 0 & 0 & 0 & 0 & 0 & 0 & 0 & 0 & 0 & 0 & 0 & 0 & 1
\end{bmatrix}$$

The matrix showing just the transient states (1, 2, 3, 5, 6, 7, 9, 10 and 11) is

$$N = \begin{bmatrix}
0 & \frac{2}{3} & 0 & \frac{1}{3} & 0 & 0 & 0 & 0 & 0 \\
0 & 0 & \frac{2}{3} & 0 & \frac{1}{3} & 0 & 0 & 0 & 0 \\
0 & 0 & 0 & 0 & 0 & \frac{1}{3} & 0 & 0 & 0 \\
0 & 0 & 0 & 0 & \frac{2}{3} & 0 & \frac{1}{3} & 0 & 0 \\
0 & 0 & 0 & 0 & 0 & \frac{2}{3} & 0 & \frac{1}{3} & 0 \\
0 & 0 & 0 & 0 & 0 & 0 & 0 & 0 & \frac{1}{3} \\
0 & 0 & 0 & 0 & 0 & 0 & 0 & \frac{2}{3} & 0 \\
0 & 0 & 0 & 0 & 0 & 0 & 0 & 0 & \frac{2}{3} \\
0 & 0 & 0 & 0 & 0 & 0 & 0 & 0 & 0
\end{bmatrix}$$

from which we can compute

$$N + N^2 + N^3 + N^4 + N^5 = \begin{bmatrix}
0 & 2/3 & 4/9 & 1/3 & 4/9 & 4/9 & 1/9 & 2/9 & 8/27 \\
0 & 0 & 2/3 & 0 & 1/3 & 4/9 & 0 & 1/9 & 2/9 \\
0 & 0 & 0 & 0 & 0 & 1/3 & 0 & 0 & 1/9 \\
0 & 0 & 0 & 0 & 2/3 & 4/9 & 1/3 & 4/9 & 4/9 \\
0 & 0 & 0 & 0 & 0 & 2/3 & 0 & 1/3 & 4/9 \\
0 & 0 & 0 & 0 & 0 & 0 & 0 & 0 & 1/3 \\
0 & 0 & 0 & 0 & 0 & 0 & 0 & 2/3 & 4/9 \\
0 & 0 & 0 & 0 & 0 & 0 & 0 & 0 & 2/3 \\
0 & 0 & 0 & 0 & 0 & 0 & 0 & 0 & 0
\end{bmatrix}$$

Adding 1 to the sum of the entries in the first row (corresponding to games beginning in State 1) gives us $\frac{104}{27} \approx 3.85$ games expected in this best-of-five series.

19. *Proof by induction.* The first statement is, "$M^1 = \begin{bmatrix} 1/2 & 0 \\ 0 & 1/2 \end{bmatrix}$," which is true by the given definition of M. Let $m \geq 2$ be given such that the first statement *not* yet proven is the one involving M^m. In particular, the previous statement,

$$M^{m-1} = \begin{bmatrix} (1/2)^{m-1} & 0 \\ 0 & (1/2)^{m-1} \end{bmatrix}$$

has already been checked to be true. In this case, we know

$$\begin{aligned}
M^m &= M \cdot M^{m-1} \\
&= \begin{bmatrix} 1/2 & 0 \\ 0 & 1/2 \end{bmatrix} \cdot \begin{bmatrix} (1/2)^{m-1} & 0 \\ 0 & (1/2)^{m-1} \end{bmatrix} \\
&= \begin{bmatrix} (1/2)^m & 0 \\ 0 & (1/2)^m \end{bmatrix}
\end{aligned}$$

which verifies the next statement, completing the induction.

21. *Proof by induction.* The first statement is, "$M^1 = \begin{bmatrix} 0 & 2(1/4)^1 \\ 2(1/4)^1 & 0 \end{bmatrix}$, and $M^2 = \begin{bmatrix} (1/4)^1 & 0 \\ 0 & (1/4)^1 \end{bmatrix}$,"
which is true by the given definition of M and the computation

$$
\begin{aligned}
M^2 &= M \cdot M \\
&= \begin{bmatrix} 0 & 1/2 \\ 1/2 & 0 \end{bmatrix} \cdot \begin{bmatrix} 0 & 1/2 \\ 1/2 & 0 \end{bmatrix} \\
&= \begin{bmatrix} (1/2)(1/2) & 0 \\ 0 & (1/2)(1/2) \end{bmatrix} \\
&= \begin{bmatrix} 1/4 & 0 \\ 0 & 1/4 \end{bmatrix}
\end{aligned}
$$

Let $m \geq 2$ be given such that the first statement *not* yet proven is the one involving M^m. In particular, the previous statement,

$$
M^{2m-3} = \begin{bmatrix} 0 & 2(1/4)^{m-1} \\ 2(1/4)^{m-1} & 0 \end{bmatrix} \text{ and}
$$

$$
M^{2m-2} = \begin{bmatrix} (1/4)^{m-1} & 0 \\ 0 & (1/4)^{m-1} \end{bmatrix}
$$

have already been checked to be true. In this case, we know

$$
\begin{aligned}
M^{2m-1} &= M \cdot M^{2m-2} \\
&= \begin{bmatrix} 0 & 1/2 \\ 1/2 & 0 \end{bmatrix} \cdot \begin{bmatrix} (1/4)^{m-1} & 0 \\ 0 & (1/4)^{m-1} \end{bmatrix} \\
&= \begin{bmatrix} 0 & (1/2)(1/4)^{m-1} \\ (1/2)(1/4)^{m-1} & 0 \end{bmatrix} \\
&= \begin{bmatrix} 0 & 2(1/4)^m \\ 2(1/4)^m & 0 \end{bmatrix}
\end{aligned}
$$

and so

$$
\begin{aligned}
M^{2m} &= M \cdot M^{2m-1} \\
&= \begin{bmatrix} 0 & 1/2 \\ 1/2 & 0 \end{bmatrix} \cdot \begin{bmatrix} 0 & 2(1/4)^m \\ 2(1/4)^m & 0 \end{bmatrix} \\
&= \begin{bmatrix} (1/4)^m & 0 \\ 0 & (1/4)^m \end{bmatrix}
\end{aligned}
$$

which verifies the next statement, completing the induction.

23. *Proof by induction.* Let M be an $n \times n$ transition matrix reflecting the one-move transition probabilities for States 1 through n of a game, and let $P(k)$ be the statement, "The entry in Row i, Column j of the matrix M^k is the probability of the game moving from State i to State j in k moves." The statement $P(1)$ is true by the definition of the transition matrix M. Let $m \geq 2$ be given such that all plays of the game consisting of less than m moves have been checked, and we are now considering m moves of the game. Let States i and j be given, and let's compute the probability that the game goes from State i to State j in exactly m moves.

In order to go from State i to State j, the game must go from State i to some State l in $m-1$ moves and then from that State l to State j in one move. By the induction hypothesis, the probability of going from State i to State l in $m-1$ moves is $M_{i,l}^{m-1}$, the entry in Row i, Column l of M^{m-1}. The probability of going from State l to State j in one move is $M_{i,j}$.

Hence by the product rule, the probability of going from State i to State l in $m-1$ moves followed by one move to State j is

$$
M_{i,l}^{k-1} \cdot M_{l,j}
$$

By the sum rule, we add all possible cases of which state is State l, to get the probability of going from State i to State j in m moves as

$$\sum_{l=1}^{n} M_{i,l}^{k-1} \cdot M_{l,j}$$

which is precisely the entry in Row i, Column j of

$$M^{k-1} \cdot M = M^k$$

This completes the induction step.

25. We use the matrix

$$N = \begin{bmatrix} 1/8 & 9/16 & 0 & 0 & 0 & 0 & 0 \\ 5/16 & 1/8 & 9/16 & 0 & 0 & 0 & 0 \\ 0 & 5/16 & 1/8 & 9/16 & 0 & 0 & 0 \\ 0 & 0 & 5/16 & 1/8 & 9/16 & 0 & 0 \\ 0 & 0 & 0 & 5/16 & 1/8 & 9/16 & 0 \\ 0 & 0 & 0 & 0 & 5/16 & 1/8 & 9/16 \\ 0 & 0 & 0 & 0 & 0 & 5/16 & 1/8 \end{bmatrix}$$

corresponding to the transient states from the matrix given in Exercise 2, and compute

$$N + N^2 + \cdots + N^{1000} \approx \begin{bmatrix} 0.765 & 1.74 & 1.70 & 1.62 & 1.49 & 1.24 & 0.797 \\ 0.967 & 1.71 & 2.64 & 2.52 & 2.31 & 1.93 & 1.24 \\ 0.524 & 1.47 & 2.17 & 3.03 & 2.77 & 2.31 & 1.49 \\ 0.278 & 0.779 & 1.68 & 2.30 & 3.03 & 2.52 & 1.62 \\ 0.142 & 0.396 & 0.855 & 1.68 & 2.17 & 2.64 & 1.70 \\ 0.0656 & 0.184 & 0.396 & 0.779 & 1.47 & 1.71 & 1.74 \\ 0.0234 & 0.0656 & 0.142 & 0.278 & 0.524 & 0.967 & 0.765 \end{bmatrix}$$

Adding 1 to the sum of the entries in the fourth row (corresponding to games beginning in State 5) gives us approximately 13.21 moves expected in this game.

27. We use the matrix

$$N = \begin{bmatrix} 0 & 1/4 & 1/4 & 1/4 & 1/4 & 0 & 0 & 0 & 0 & 0 & 0 & 0 & 0 & 0 & 0 \\ 0 & 1/4 & 0 & 0 & 0 & 1/4 & 1/4 & 1/4 & 0 & 0 & 0 & 0 & 0 & 0 & 0 \\ 0 & 0 & 1/4 & 0 & 0 & 1/4 & 0 & 0 & 1/4 & 1/4 & 0 & 0 & 0 & 0 & 0 \\ 0 & 0 & 0 & 1/4 & 0 & 0 & 1/4 & 0 & 1/4 & 0 & 1/4 & 0 & 0 & 0 & 0 \\ 0 & 0 & 0 & 0 & 1/4 & 0 & 0 & 1/4 & 0 & 1/4 & 1/4 & 0 & 0 & 0 & 0 \\ 0 & 0 & 0 & 0 & 0 & 1/2 & 0 & 0 & 0 & 0 & 0 & 1/4 & 1/4 & 0 & 0 \\ 0 & 0 & 0 & 0 & 0 & 0 & 1/2 & 0 & 0 & 0 & 0 & 1/4 & 0 & 1/4 & 0 \\ 0 & 0 & 0 & 0 & 0 & 0 & 0 & 1/2 & 0 & 0 & 0 & 0 & 1/4 & 1/4 & 0 \\ 0 & 0 & 0 & 0 & 0 & 0 & 0 & 0 & 1/2 & 0 & 0 & 1/4 & 0 & 0 & 1/4 \\ 0 & 0 & 0 & 0 & 0 & 0 & 0 & 0 & 0 & 1/2 & 0 & 0 & 1/4 & 0 & 1/4 \\ 0 & 0 & 0 & 0 & 0 & 0 & 0 & 0 & 0 & 0 & 1/2 & 0 & 0 & 1/4 & 1/4 \\ 0 & 0 & 0 & 0 & 0 & 0 & 0 & 0 & 0 & 0 & 0 & 3/4 & 0 & 0 & 0 \\ 0 & 0 & 0 & 0 & 0 & 0 & 0 & 0 & 0 & 0 & 0 & 0 & 3/4 & 0 & 0 \\ 0 & 0 & 0 & 0 & 0 & 0 & 0 & 0 & 0 & 0 & 0 & 0 & 0 & 3/4 & 0 \\ 0 & 0 & 0 & 0 & 0 & 0 & 0 & 0 & 0 & 0 & 0 & 0 & 0 & 0 & 3/4 \end{bmatrix}$$

corresponding to the transient states from the matrix given in Exercise 4, and compute $N + N^2 + \cdots + N^{1000} \approx$

$$\begin{bmatrix}
0.0 & 0.33 & 0.33 & 0.33 & 0.33 & 0.33 & 0.33 & 0.33 & 0.33 & 0.33 & 0.33 & 1.0 & 1.0 & 1.0 & 1.0 \\
0.0 & 0.33 & 0.0 & 0.0 & 0.0 & 0.67 & 0.67 & 0.67 & 0.0 & 0.0 & 0.0 & 1.33 & 1.33 & 1.33 & 0.0 \\
0.0 & 0.0 & 0.33 & 0.0 & 0.0 & 0.67 & 0.0 & 0.0 & 0.67 & 0.67 & 0.0 & 1.33 & 1.33 & 0.0 & 1.33 \\
0.0 & 0.0 & 0.0 & 0.33 & 0.0 & 0.0 & 0.67 & 0.0 & 0.67 & 0.0 & 0.67 & 1.33 & 0.0 & 1.33 & 1.33 \\
0.0 & 0.0 & 0.0 & 0.0 & 0.33 & 0.0 & 0.0 & 0.67 & 0.0 & 0.67 & 0.67 & 0.0 & 1.33 & 1.33 & 1.33 \\
0.0 & 0.0 & 0.0 & 0.0 & 0.0 & 1.0 & 0.0 & 0.0 & 0.0 & 0.0 & 0.0 & 2.0 & 2.0 & 0.0 & 0.0 \\
0.0 & 0.0 & 0.0 & 0.0 & 0.0 & 0.0 & 1.0 & 0.0 & 0.0 & 0.0 & 0.0 & 2.0 & 0.0 & 2.0 & 0.0 \\
0.0 & 0.0 & 0.0 & 0.0 & 0.0 & 0.0 & 0.0 & 1.0 & 0.0 & 0.0 & 0.0 & 0.0 & 2.0 & 2.0 & 0.0 \\
0.0 & 0.0 & 0.0 & 0.0 & 0.0 & 0.0 & 0.0 & 0.0 & 1.0 & 0.0 & 0.0 & 2.0 & 0.0 & 0.0 & 2.0 \\
0.0 & 0.0 & 0.0 & 0.0 & 0.0 & 0.0 & 0.0 & 0.0 & 0.0 & 1.0 & 0.0 & 0.0 & 2.0 & 0.0 & 2.0 \\
0.0 & 0.0 & 0.0 & 0.0 & 0.0 & 0.0 & 0.0 & 0.0 & 0.0 & 0.0 & 1.0 & 0.0 & 0.0 & 2.0 & 2.0 \\
0.0 & 0.0 & 0.0 & 0.0 & 0.0 & 0.0 & 0.0 & 0.0 & 0.0 & 0.0 & 0.0 & 3.0 & 0.0 & 0.0 & 0.0 \\
0.0 & 0.0 & 0.0 & 0.0 & 0.0 & 0.0 & 0.0 & 0.0 & 0.0 & 0.0 & 0.0 & 0.0 & 3.0 & 0.0 & 0.0 \\
0.0 & 0.0 & 0.0 & 0.0 & 0.0 & 0.0 & 0.0 & 0.0 & 0.0 & 0.0 & 0.0 & 0.0 & 0.0 & 3.0 & 0.0 \\
0.0 & 0.0 & 0.0 & 0.0 & 0.0 & 0.0 & 0.0 & 0.0 & 0.0 & 0.0 & 0.0 & 0.0 & 0.0 & 0.0 & 3.0
\end{bmatrix}$$

Adding 1 to the sum of the entries in the first row (corresponding to games beginning in State 1) gives us approximately 8.33 moves expected in this game.

$$M = \begin{bmatrix}
0 & 1/4 & 1/2 & 1/4 & 0 & 0 \\
1/4 & 0 & 1/4 & 1/4 & 1/4 & 0 \\
1/4 & 0 & 0 & 1/2 & 1/4 & 0 \\
1/4 & 0 & 0 & 1/4 & 1/4 & 1/4 \\
0 & 0 & 0 & 1/4 & 0 & 3/4 \\
0 & 0 & 0 & 0 & 0 & 1
\end{bmatrix}$$

29. We use the matrix

$$N = \begin{bmatrix} 5/6 \end{bmatrix}$$

corresponding to the transient states from the matrix given in Exercise 6, and compute

$$N + N^2 + \cdots + N^{1000} \approx \begin{bmatrix} 5.0 \end{bmatrix}$$

Adding 1 to the sum of the entries in the first row (corresponding to games beginning in State 1) gives us approximately 6.0 moves expected in this game.

Section 7.1 exercises

1. (a) Nine nodes and eleven edges.

(c) The sum of the degrees is 22, which is twice the number of edges, 11.

(d) *ABDA, DEFIHGD, FIHF, DEFHGD*

4. (a) 12346781 and 12643781

(c) 3487645

5. (a) $d(1) = 4, d(2) = 4, d(3) = 2, d(4) = 2, d(5) = 6, d(6) = 6, d(7) = 2, d(8) = 2, d(9) = 4, d(10) = 4$

(c) One possibility is $1, 3, 5, 2, 6, 10, 5, 7, 9, 6, 1$

7. (a) Here are two possibilities:

(c) Here are two possibilities:

(d) If there are 7 nodes in a simple graph, the largest degree is six.

(e) There cannot be an odd number of vertices of odd degree.

9. (a) $3, 4, 8, 2, 6, 4, 5, 6, 7, 8, 1, 2, 3, 7$ is an Eulerian trail.

(b) There are four vertices of odd degree, so this graph has no Eulerian circuit or trail.

(c) $1, 2, 3, 4, 5, 1, 6, 2, 7, 3, 8, 4, 9, 5, 10, 6, 7, 8, 9, 10, 1$ is an Eulerian circuit.

(d) $1, 2, 4, 6, 8, 10, 9, 7, 5, 3, 1, 5, 2, 6, 10, 5, 9, 6, 1$ is an Eulerian circuit.

11. In this graph, there is an Eulerian trail since the "In" and "Out" nodes are the only ones with odd degree.

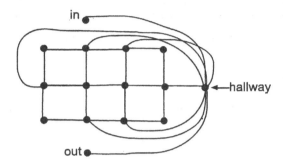

12. (a) Here is the only connected graph with 0 edges with every vertex having even degree. The trivial circuit a is an Eulerian circuit.

(b) Here is the only connected graph with 1 edge with every vertex having even degree. The circuit $a, 1, a$ is an Eulerian circuit.

(c) Here are both connected graphs with 2 edges with every vertex having even degree. The graph on the left has the Eulerian circuit $a, 1, a, 2, a$, and the graph on the right has the Eulerian circuit $a, 1, b, 2, a$.

(d) Here are all connected graphs G with 3 edges with every vertex having even degree. The first graph has the Eulerian circuit $a, 1, a, 2, a, 3, a$, the second graph has the Eulerian circuit $a, 1, a, 2, b, 3, a$, and the third graph has the Eulerian circuit $a, 1, b, 3, c, 2, a$.

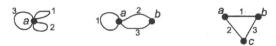

13. Yes, it is possible. Sarah can walk 1,6,10 and Emily 4,2,3,4,5,6,9,10,7,1,3,5,9,8,7,4,8,5, for example. This will be possible in any connected graph with no more than four nodes of odd degree. The graph below cannot be covered in this way.

15. a) 2   b) 0   c) 2

17. Define the graph G having nodes labeled $0, 1, 2, 3, 4, 5, 6$ (for the number of spots on the dominoes) and an edge for each domino with endpoints reflecting the number of spots on that domino. In this graph, all nodes have degree 6. When one domino (edge) is removed, the nodes labeled with the two numbers on that domino have degree 5. The resulting graph has an Eulerian path starting and ending at these two nodes of odd degree. The trick works because of this.

19. A graph like this would have 10 vertices and $(3 + 3 + 2 + 2 + 1 + 1 + 1 + 1 + 1)/2 = 8$ edges. By the previous exercise, simple graph like this cannot be connected.

21. If you have one vertex with no edges and $n - 1$ vertices with all possible edges between them, then you will have a disconnected graph with $C(n - 1, 2) = \frac{(n-1)(n-2)}{2}$ edges. This is the most possible.

23. The connected graph has exactly two nodes of odd degree, and these two nodes have an edge connecting them.

Section 7.2 exercises

1. Refer to the three graphs in the figure below.

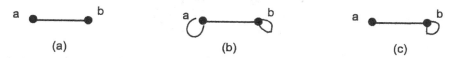

(a)   (b)   (c)

(a) The walk $W = a, 1, b, 1, a$ is closed but does not contain a cycle.

(b) Every vertex is of degree 3 but the only cycles have size 1.

(c) There are two vertices, one of degree 1 and one of degree 3.

2. The three blanks should be filled in as follows: G is connected; W itself; W with the edge $[v_0, v_1]$ replaced by the walk $v_0, v_n, \ldots, v_2, v_1$

3. (a) $1, 2, 4$

(c) $9, 3, 2, 4$

(e) $3, 5, 4, 8, 9, 3, 2, 8, 7 \rightarrow 3, 2, 8, 7$

7. *Proof.* Let G be a simple graph with n vertices, where $n \geq 2$. Since G is simple, the degree of each vertex must be no more than $n - 1$. Moreover, if there is a vertex of degree $n - 1$, then that vertex is adjacent to every other vertex in the graph, making it impossible for there to also be a degree 0 vertex. Create boxes labeled 1 to $n - 2$ and another box labeled "0 or $n - 1$," and assign each vertex of the graph to a box based on its degree. Since there are $n - 1$ boxes and n vertices, the Pigeonhole Principle tells us that at least one box contains two vertices. These two vertices must have the same degree.

9. *Proof.* Let G be a graph on n nodes, and let P be a path in G. Since a path cannot use the same node twice, P contains no more than n nodes. Since for every path the number of edges is one less than the number of nodes, the number of edges in P cannot be more than $n - 1$.

10. The contrapositive statement is, "For every connected graph G, if there exists a pair of vertices a, b in G with two (or more) paths from a to b, then G has at least one cycle.

Proof. Let G be a connected graph with vertices a and b such that there are two paths from a to b. Label the paths $a = v_0, v_1, v_2, \ldots, v_n = b$ and $a = w_0, w_1, w_2, \ldots, w_m = b$. Let i be the smallest integer for which the edge $[v_i, v_{i+1}]$ from the first path is not in the second path. (This must happen since the paths are different.) In this case, v_i is a vertex, say w_k, in the second path since otherwise there would be an earlier mentioned edge in the first path not in the second. Let $j > i$ be the smallest integer for which vertex v_j is in the second path as say w_l. (This must happen since both paths come together at b for sure. They might come together for the first time earlier though.) If we have done all of this, then following the first path from v_i to $v_j = w_l$ and then the second path from w_l to $w_k = v_i$ will form a cycle.

11. Let G be a tree with at least one edge, let $e = [v_0, v_1]$ denote the deleted edge, and let G' denote the resulting graph. By the previous exercise, the only path from v_0 to v_1 in G is v_0 to v_1. Thus in G' there is no path from v_0 to v_1, and hence no walk from v_0 to v_1.

Define $H_0 = \{v \in V \mid \text{there is a walk from } v \text{ to } v_0 \text{ in } G'\}$ and $H_1 = \{v \in V \mid \text{there is a walk from } v \text{ to } v_1$ in $G'\}$. Since for every vertex v in G there is a walk in G from v to v_0 and a walk in G from v to v_1, it follows that H_0 and H_1 are each connected and that every vertex is in one of the two sets. Also note that no vertex can be in both sets since otherwise we could build a path from v_0 to v_1 in G'. It follows that H_0 and H_1 are the connected components of G'.

14. The blanks should be filled in as follows: Any tree G with 0 edges has one node; The graph $\bullet$ has one vertex and no edges; Proposition 4; $m - 1$; $m - 1$; m; G has one more node that G'; $m + 1$.

17. All relevant graphs with no edges, one edge, two edges or three edges are given in the figure below.

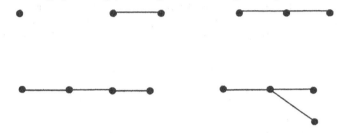

20. *Proof by induction.* Let $P(k)$ stand for, "A simple graph with n vertices, k connected components, and no cycles has $n - k$ edges." The first statement $P(1)$ is true by Theorem 7 since if $k = 1$, such a graph is connected and therefore a tree.

Now let $m > 1$ be given such that statements $P(1), \ldots, P(m-1)$ have been checked. Let G be a graph with n vertices, m connected components, and no cycles. Choose any component H, and let G' be the graph with H removed. Let n_1 be the number of vertices in H, and let n_2 be the number of vertices in G'. Observe that $n_1 + n_2 = n$.

By $P(1)$, the component H has $n_1 - 1$ edges, and by statement $P(m-1)$, the graph G' has $n_2 - (m-1)$ edges. Thus G has a total of

$$(n_1 - 1) + (n_2 - (m - 1)) = n_1 - 1 + n_2 - m + 1 = n_1 + n_2 - m = n - m$$

edges. This establishes statement $P(m)$, and the result follows by induction.

21. The blanks should be filled in as follows: Exercise 20; 1; $n - 1$; 1; $-n$; $\frac{n^2 - n}{2}$; $\frac{n}{2}$; 1; $n - 1$; $\frac{(n-1)(n-2)}{2}$.

23. (a) Using Prim's algorithm, there is only one, with edges $[a, d], [a, c], [a, b]$ and a total weight of 6.

   (b) Using Prim's algorithm, there in only one, with edges $[a, e], [a, d], [a, b], [c, e]$ and a total weight of 13.

   (c) Using Prim's algorithm, there is only one, with edges $[a, b], [a, d], [a, e], [c, e]$ and a total weight of 18.

25. (a) We list the edges in the order $[a, d], [a, c], [a, b], [b, d], [c, d], [b, c]$. Edges are added at steps $1, 2$ and 3, giving the minimal spanning tree with edges $[a, d], [a, c], [a, b]$ and total weight 6.

   (b) The algorithm adds the edges $[a, e], [a, d], [a, b], [c, e]$ for a total weight of 13.

   (c) The algorithm adds $[a, b]$ and $[a, d]$ in the first two steps, and $[a, e]$ and $[c, e]$ in the next two steps.

26. *Proof by induction.* Let $P(n)$ be the statement, "T_n is included in a minimal spanning tree of G." Since T_0 has the same vertex set as G and no edges, then clearly *any* minimal spanning tree of G will include T_0 as a subgraph, hence statement $P(0)$ is true.

Let $m \geq 1$ be given such that statements $P(0), \ldots, P(m-1)$ have all been checked to be true. Now we consider statement $P(m)$. According to the algorithm, T_m was formed by adding the edge e_m to the tree T_{m-1}. By the inductive hypothesis $P(m-1)$, we know that T_{m-1} is a subgraph of a minimal spanning tree of G. Let's call this minimal spanning tree $\mathbf{T}$. The edge e_m either is in $\mathbf{T}$ or it is not, so we argue by cases.

Case 1. If it so happens that the edge e_m is actually in **T**, then it must be the case that T_m is a subgraph of **T**, which means that statement $P(m)$ is true.

Case 2. If e_m is *not* an edge in **T**, then we can form the new graph H by adding e_m to **T**. By Exercise 12, this new graph will have exactly one cycle. Let f be the edge on this cycle with the smallest possible weight (and so the weight of f is no more than the weight of e_m), and let **T'** be the graph formed by removing f from H. By Exercise 13, **T'** is a spanning tree of G. Now consider the two possibilities for how the weight of f compares to the weight of e_m.

> **Case 2a.** If the weight of f is equal to the weight of e_m, then **T'** has the same weight as **T**, and hence **T'** is a minimal spanning tree of G.

> **Case 2b.** If the weight of f is less than the weight of e_m, then f must have appeared earlier in the list of edges that was pre-sorted from smallest to largest weight. That is, $f = e_k$ for some $k < m$. Since f is not in T_m, then it is also not in T_k, which according to the algorithm can only be the case if adding f to T_{k-1} creates a cycle. But in this case, adding f to T_{m-1} will create the same cycle, which contradicts the fact that **T'** is a tree. Hence, this case is impossible.

Hence, in the only possible case, T_m is included in the minimal spanning tree **T'** of G. This establishes statement $P(m)$ in Case 2 as well, and this completes the induction.

27. The blanks should be filled in as follows: $n + 1$; $2(n + 1) = 2n + 2$; $2n$; $2n + 2 > 2n$.

29. Using Kruskal's algorithm, we add the edges in this order: ae, fj, ld, bc, bf, gj, ij, im, kl, ab, hi, jn, ad. The total cost is $3(\$100) + 6(\$200) + 3(\$300) + \$400 = \$2800$.

31. (a) One solution is to remove cd from $acda$, then bd from $abda$, and finally bc from $abca$. This leaves the spanning tree ab, ac, ad with a total weight of 6.

(b) One solution is to remove bc from $abcdea$, then bc from $abdea$, then be from $abea$, then ac from $acea$, then de from $adea$, and finally cd from $adcea$. This leaves the spanning tree ab, ad, ae, ce with a total weight of 13.

(c) One solution is to remove bc from $abca$, then de from $cdec$, then cd from $acda$, then ac from $acea$, then bd from $abda$, and finally be from $abea$. This leaves the spanning tree ab, ad, ae, ce with a total weight of 18.

33. *Proof by induction.* Let $P(n)$ denote the statement, "In the algorithm of Exercise 30, graph G_n contains a minimal spanning tree of G." Since G_0 is G and G is given as being connected, $P(0)$ is true. Now let $m \geq 1$ be given such that statements $P(0), \ldots, P(m - 1)$ have all been checked to be true. Recall that graph G_m is formed from G_{m-1} be removing the most expensive edge e from some cycle C in G_{m-1}. By statement $P(m-1)$, graph G_{m-1} contains a spanning tree T that is a minimal spanning tree for G. Since trees don't have cycles, the cycle C must include some edge f that is not in T. There are two cases to consider:

Case 1. If $e = f$, then all edges of T are included in G_m, so G_m contains the minimal spanning tree T of G.

Case 2. If $e \neq f$, then edge e must weigh at least as much edge f (since the algorithm deleted e to form G_m), so the new tree T' formed by adding f to T and removing e will weigh no more overall than T. Hence T' is a different minimal spanning tree for G, and T' is contained in G_m.

Since in either case, G_m contains a minimal spanning tree for G, this establishes statement $P(m)$, completing the induction.

Section 7.3 exercises

1. (a) Graph G_2 has a nodes of degree 4 and graph G_1 does not.

(b) Graph H_1 has a cycle of length 5 and graph H_2 does not.

3. If we write down the degrees of each node, we find that graph (B) has two nodes of degree 2, and the others have only one. Hence (B) is not isomorphic to any of the others. Since all the graphs have a unique node of degree 1, these would have to correspond to each other under an isomorphism. Note that the degree 1 node in graph (D) is adjacent to a degree 3 node, while in graphs (A) and (C), the degree 1 node is adjacent to a degree 4 node. Thus (D) is not isomorphic to (A) or (C). This leaves (A) and (C) as the only candidates. To see the isomorphism, imagine moving the bottom left node in (C) halfway toward the bottom right node, then dragging the node in the middle in (C) down to the bottom left.

5. The graph on the left has no cycles of length 3 while the graph on the right does. Another difference is that the graph on the left is $K_{3,3}$ so it is non-planar, while the graph on the right has no edge-crossings.

 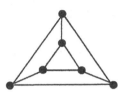

7. The three graphs are $K_{5,5}$ and the two graphs shown in the figure below.

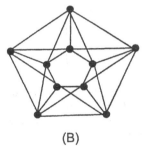 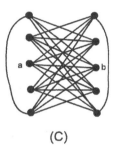

(B) (C)

All cycles in $K_{5,5}$ have an even number of edges. Graphs (B) and (C) contain cycles of length 3, so neither is isomorphic to $K_{5,5}$. Graph (B) has the property that every edge is on some cycle of length 3. In graph (C), edge $[a, b]$ is not on such a cycle, hence graphs (B) and (C) are not isomporphic.

10. (a)

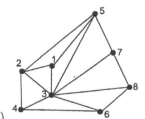

(b)

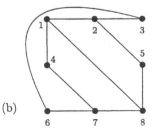

11. (a) Suppose (A) is planar. (A) has 8 vertices, 24 edges, every edge is on a cycle, and the smallest cycle is length 3. By Euler's formula, (A) has exactly $24 + 2 - 8 = 18$ faces. By Theorem 7, the number of edges is at least $\frac{3}{2}$ times the number of faces, that is $24 \geq \frac{3}{2} \cdot 18$, or $24 \geq 27$. Since this is a contradiction, we conclude that (A) is not planar.

(b) Suppose (B) is planar. (B) has 8 vertices, 16 edges, every edge is on a cycle, and the smallest cycle is length 4. By Euler's formula, (B) has exactly $16 + 2 - 8 = 10$ faces. By Theorem 7, the number of edges is at least $\frac{4}{2}$ times the number of faces, that is $16 \geq \frac{4}{2} \cdot 10$, or $16 \geq 20$. Since this is a contradiction, we conclude that (B) is not planar.

(c) Suppose (C) is planar. (C) has 10 vertices, 15 edges, every edge is on a cycle, and the smallest cycle is length 5. By Euler's formula, (C) has exactly $15 + 2 - 8 = 9$ faces. By Theorem 7, the number of edges is at least $\frac{5}{2}$ times the number of faces, that is $15 \geq \frac{5}{2} \cdot 9$, or $15 \geq 22.5$. Since this is a contradiction, we conclude that (A) is not planar.

12. (a) This graph is non-planar because it contains $K_{3,3}$. Every node in the set $\{1, 2, 6\}$ is connected to every node in $\{3, 4, 5\}$. The additional edges $[4, 5]$ and $[3, 4]$ only make it worse.

(b) The embedding for this planar graph is shown below:

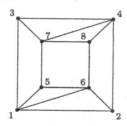

(c) This graph has 8 vertices and 20 edges, so by Euler's formula it will have 14 faces if it *is* planar. However, every edge is on a cycle and the minimum length of a cycle is 3, so Theorem 7 tells us that if the graph is planar, then the number of edges must be $\frac{3}{2}$ times the number of faces. Since $20 < \frac{3}{2}(14)$, we conclude that the graph is not planar.

13. (a) For the illustration, $V = 12, E = 18, F = 8$ and $12 + 8 = 18 + 2$. In general, $V = 2k, E = 3k, F = k + 2; 2k + (k + 2) = 3k + 2$.

(c) For the illustration, $V = 5, E = 8, F = 5$ and $5 + 5 = 8 + 2$. In general, $V = k + 1, E = 2k, F = k + 1; (k + 1) + (k + 1) = 2k + 2$.

15. *Proof.* By Theorem 7, $m \geq \frac{3}{2}F$, hence $F \leq \frac{2}{3}m$. By Euler's formula, $n + F = m + 2$. So $m + 2 = n + F \leq n + \frac{2}{3}m$. Solving the inequality $m + 2 \leq n + \frac{2}{3}m$, we get $m \leq 3n - 6$.

17. In the figure below, the dashed line indicates a line on the opposite side of the surface. Each of the houses A, B and C are connected to every utility 1, 2 and 3.

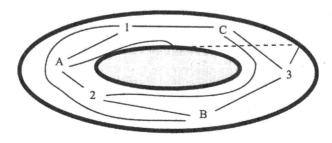

Section 7.4 exercises

1. (a) $$\begin{bmatrix} 0 & 1 & 0 & 1 \\ 1 & 0 & 1 & 1 \\ 0 & 1 & 0 & 0 \\ 1 & 1 & 0 & 0 \end{bmatrix}$$

(b) $$\begin{bmatrix} 0 & 1 & 1 & 1 \\ 1 & 0 & 1 & 1 \\ 1 & 1 & 1 & 0 \\ 1 & 1 & 0 & 1 \end{bmatrix}$$

(c) $$\begin{bmatrix} 0 & 1 & 0 & 1 & 1 \\ 0 & 0 & 0 & 0 & 0 \\ 0 & 0 & 1 & 1 & 0 \\ 0 & 1 & 0 & 0 & 0 \\ 0 & 0 & 0 & 1 & 0 \end{bmatrix}$$

(d) $$\begin{bmatrix} 1 & 1 & 1 \\ 1 & 1 & 1 \\ 1 & 1 & 1 \end{bmatrix}$$

(e) $$\begin{bmatrix} 0 & 1 & 0 & 0 & 0 \\ 1 & 0 & 0 & 0 & 0 \\ 0 & 0 & 0 & 0 & 0 \\ 0 & 0 & 1 & 0 & 1 \\ 0 & 0 & 1 & 0 & 0 \end{bmatrix}$$

3. (a) The degree of the ith node is the sum of the ith row.

(b) The degree of the ith node is the sum of the ith row plus the value of the (i,i) entry. (This means the (i,i) entry is counted twice.)

4. (a) The outdegree for the ith node is the sum of the ith row. The indegree is the sum of the ith column.

5. (a) $$M = \begin{bmatrix} 0 & 0 & 1 & 1 & 0 & 0 \\ 1 & 0 & 0 & 0 & 1 & 0 \\ 0 & 1 & 0 & 1 & 0 & 0 \\ 0 & 1 & 0 & 0 & 1 & 1 \\ 1 & 0 & 1 & 0 & 0 & 1 \\ 1 & 1 & 1 & 0 & 0 & 0 \end{bmatrix}$$

$$M^2 = \begin{bmatrix} 0 & 2 & 0 & 1 & 1 & 1 \\ 1 & 0 & 2 & 1 & 0 & 1 \\ 1 & 1 & 0 & 0 & 2 & 1 \\ 3 & 1 & 2 & 0 & 1 & 1 \\ 1 & 2 & 2 & 2 & 0 & 0 \\ 1 & 1 & 1 & 2 & 1 & 0 \end{bmatrix}$$

$$M^3 = \begin{bmatrix} 4 & 2 & 2 & 0 & 3 & 2 \\ 1 & 4 & 2 & 3 & 1 & 1 \\ 4 & 1 & 4 & 1 & 1 & 2 \\ 3 & 3 & 5 & 5 & 1 & 1 \\ 2 & 4 & 1 & 3 & 4 & 2 \\ 2 & 3 & 2 & 2 & 3 & 3 \end{bmatrix}$$

$$M + M^2 + M^3 = \begin{bmatrix} 4 & 4 & 3 & 2 & 4 & 3 \\ 3 & 4 & 4 & 4 & 2 & 2 \\ 5 & 3 & 4 & 2 & 3 & 3 \\ 6 & 5 & 7 & 5 & 3 & 3 \\ 4 & 6 & 4 & 5 & 4 & 3 \\ 4 & 5 & 4 & 4 & 4 & 3 \end{bmatrix}$$

So there are three walks of length 3 or less from node 1 to node 6. They are $1 - 4 - 6$, $1 - 3 - 4 - 6$ and $1 - 4 - 5 - 6$.

(b) $M = \begin{bmatrix} 0 & 1 & 1 & 1 & 1 & 1 \\ 0 & 0 & 1 & 1 & 1 & 1 \\ 0 & 0 & 0 & 1 & 1 & 1 \\ 0 & 0 & 0 & 0 & 1 & 1 \\ 0 & 0 & 0 & 0 & 0 & 1 \\ 0 & 0 & 0 & 0 & 0 & 0 \end{bmatrix}$

$M^2 = \begin{bmatrix} 0 & 0 & 1 & 2 & 3 & 4 \\ 0 & 0 & 0 & 1 & 2 & 3 \\ 0 & 0 & 0 & 0 & 1 & 2 \\ 0 & 0 & 0 & 0 & 0 & 1 \\ 0 & 0 & 0 & 0 & 0 & 0 \\ 0 & 0 & 0 & 0 & 0 & 0 \end{bmatrix}$

$M^3 = \begin{bmatrix} 0 & 0 & 0 & 1 & 3 & 6 \\ 0 & 0 & 0 & 0 & 1 & 3 \\ 0 & 0 & 0 & 0 & 0 & 1 \\ 0 & 0 & 0 & 0 & 0 & 0 \\ 0 & 0 & 0 & 0 & 0 & 0 \\ 0 & 0 & 0 & 0 & 0 & 0 \end{bmatrix}$

$M + M^2 + M^3 = \begin{bmatrix} 0 & 1 & 2 & 4 & 7 & 11 \\ 0 & 0 & 1 & 2 & 4 & 7 \\ 0 & 0 & 0 & 1 & 2 & 4 \\ 0 & 0 & 0 & 0 & 1 & 2 \\ 0 & 0 & 0 & 0 & 0 & 1 \\ 0 & 0 & 0 & 0 & 0 & 0 \end{bmatrix}$

So there are eleven walks of length 3 or less from 1 to 6: $1-6$, $1-2-6$, $1-3-6$, $1-4-6$, $1-5-6$, $1-2-3-6$, $1-2-4-6$, $1-2-5-6$, $1-3-4-6$, $1-3-5-6$ and $1-4-5-6$.

(c) $M = \begin{bmatrix} 0 & 1 & 1 & 1 & 1 & 1 \\ 0 & 0 & 0 & 1 & 1 & 1 \\ 0 & 0 & 0 & 0 & 1 & 1 \\ 0 & 0 & 0 & 0 & 1 & 0 \\ 0 & 0 & 0 & 0 & 0 & 0 \\ 0 & 0 & 0 & 0 & 1 & 0 \end{bmatrix}$

$M^2 = \begin{bmatrix} 0 & 0 & 0 & 1 & 4 & 2 \\ 0 & 0 & 0 & 0 & 2 & 0 \\ 0 & 0 & 0 & 0 & 1 & 0 \\ 0 & 0 & 0 & 0 & 0 & 0 \\ 0 & 0 & 0 & 0 & 0 & 0 \\ 0 & 0 & 0 & 0 & 0 & 0 \end{bmatrix}$

$M^3 = \begin{bmatrix} 0 & 0 & 0 & 0 & 3 & 0 \\ 0 & 0 & 0 & 0 & 0 & 0 \\ 0 & 0 & 0 & 0 & 0 & 0 \\ 0 & 0 & 0 & 0 & 0 & 0 \\ 0 & 0 & 0 & 0 & 0 & 0 \\ 0 & 0 & 0 & 0 & 0 & 0 \end{bmatrix}$

$M + M^2 + M^3 = \begin{bmatrix} 0 & 1 & 1 & 2 & 8 & 3 \\ 0 & 0 & 0 & 1 & 3 & 1 \\ 0 & 0 & 0 & 0 & 2 & 1 \\ 0 & 0 & 0 & 0 & 1 & 0 \\ 0 & 0 & 0 & 0 & 0 & 0 \\ 0 & 0 & 0 & 0 & 1 & 0 \end{bmatrix}$

So there are three walks of length 3 or less from 1 to 6: $1-6$, $1-2-6$ and 1-3-6.

6. (a) $\{(1,3),(1,4),(2,1),(2,5),(3,2),(3,4),(4,2),(4,5),(4,6),(5,1),(5,3),(5,6),(6,1),(6,2),(6,3)\}$

104

(b) $\{(1,2),(1,3),(1,4),(1,5),(1,6),(2,3),(2,4),(2,5),(2,6),(3,4),(3,5),(3,6),(4,5),(4,6),(5,6)\}$

(c) $\{(1,2),(1,3),(1,4),(1,5),(1,6),(2,4),(2,5),(2,6),(3,5),(3,6),(4,5),(6,5)\}$

7. (a) The three matrices are
$$\begin{bmatrix} 1 & 1 & 1 & 0 & 1 & 1 \\ 1 & 1 & 1 & 1 & 1 & 1 \\ 1 & 1 & 1 & 1 & 1 & 1 \\ 1 & 1 & 1 & 1 & 1 & 1 \\ 1 & 1 & 1 & 1 & 1 & 1 \\ 1 & 1 & 1 & 1 & 1 & 1 \end{bmatrix}, \begin{bmatrix} 0 & 0 & 0 & 1 & 1 & 1 \\ 0 & 0 & 0 & 0 & 1 & 1 \\ 0 & 0 & 0 & 0 & 0 & 1 \\ 0 & 0 & 0 & 0 & 0 & 0 \\ 0 & 0 & 0 & 0 & 0 & 0 \\ 0 & 0 & 0 & 0 & 0 & 0 \end{bmatrix} \text{ and } \begin{bmatrix} 0 & 0 & 0 & 0 & 1 & 0 \\ 0 & 0 & 0 & 0 & 0 & 0 \\ 0 & 0 & 0 & 0 & 0 & 0 \\ 0 & 0 & 0 & 0 & 0 & 0 \\ 0 & 0 & 0 & 0 & 0 & 0 \\ 0 & 0 & 0 & 0 & 0 & 0 \end{bmatrix}$$

(b) Let $A = \{1,2,3,4,5,6\}$. For the first graph, $R \circ R \circ R$ is $(A \times A) - \{1,4\}$. For the second graph, $R \circ R \circ R = \{(1,4),(1,5),(1,6),(2,5),(2,6),(3,6)\}$. For the third graph, $R \circ R \circ R = \{(1,5)\}$.

8. (a) The matrices for R and $R \circ R$, respectively, are shown below, followed by the graphs of each.

$$\begin{bmatrix} 0 & 1 & 0 & 0 & 0 & 0 \\ 0 & 0 & 1 & 0 & 0 & 0 \\ 0 & 0 & 0 & 1 & 0 & 0 \\ 0 & 0 & 0 & 0 & 1 & 0 \\ 0 & 0 & 0 & 0 & 0 & 1 \\ 0 & 0 & 0 & 0 & 0 & 0 \end{bmatrix} \qquad \begin{bmatrix} 0 & 0 & 1 & 0 & 0 & 0 \\ 0 & 0 & 0 & 1 & 0 & 0 \\ 0 & 0 & 0 & 0 & 1 & 0 \\ 0 & 0 & 0 & 0 & 0 & 1 \\ 0 & 0 & 0 & 0 & 0 & 0 \\ 0 & 0 & 0 & 0 & 0 & 0 \end{bmatrix}$$

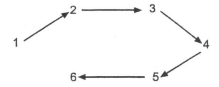

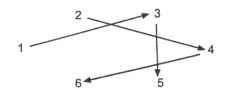

9. (a) It determines all pairs of nodes between which there is a walk of length 5 or less.

(c) $M^{(1)} \vee M^{(2)} \vee M^{(3)} \vee M^{(4)} \vee M^{(5)} \vee M^{(6)} \vee M^{(7)} \vee M^{(8)} \vee M^{(9)}$

10. (a) $2 \cdot 5 = 10$

(c) $1 \cdot 2 = 2$

(e) $2 \cdot 5 + 3 \cdot 2 + 1 \cdot 0 + 4 \cdot 3 + 0 \cdot 2 + 2 \cdot 0 + 0 \cdot 4 + 1 \cdot 2 + 0 \cdot 7 = 10 + 6 + 0 + 12 + 0 + 0 + 0 + 2 + 0 = 30$

11. (a) There are $a_{61} \cdot b_{13}$ walks from 6 to 3 that begin by going to node 1, $a_{62} \cdot b_{23}$ that begin by going to node 2, and so on, for a total of $\sum_{t=1}^{9} a_{6,t} \cdot b_{t,3}$

(c) $\sum_{t=1}^{9} a_{i,t} \cdot b_{t,j}$

12. (a) For every pair of integers i and j, the Row i, Column j entry of M^1 counts the number of 1-step walks from node i to node j.

(b) M^1 is the same as M. The Row i, Column j entry of M is the number of edges from node i to node j, and an edge is the same as a 1-step walk.

(c) For every pair of integers i and j, the Row i, Column j entry of M^{k-1} counts the number of $(k-1)$-step walks from node i to node j.

13. Let $P(k)$ be the statement, "For every pair of integers i and j, the Row i, Column j entry of M^k counts the number of k-step walks from node i to node j." In Exercise 13(b), we established $P(1)$. Now let $m \geq 2$ be given such that statements $P(1), \ldots, P(m-1)$ have been established. To establish statement $P(m)$, let i and j be given. We will show that the Row i, Column j entry of M^m counts the number of m-step walks from node i to node j. Let C_t be the count of the m-step walks from node i to node j whose first edge goes from node i to node t. Clearly, there are $\sum_{t=1}^{n} C_t$ m-step walks from node i to node j, since every such walk must contain a first edge leading to exactly one of the nodes in the graph. But C_t, by the product rule of counting, may be calculated as the product of these two counts:

- the number of 1-step walks from node i to node t
- the number of $(m-1)$-step walks from node t to node j. By $P(1)$, the former is the Row i, Column t entry of M. By $P(m-1)$ the latter is the Row t, Column j entry of M^{m-1}.

Thus $C_t = M_{i,t}(M^{m-1})_{t,j}$, and the total count is the sum of these expressions, $\sum_{i=1}^{n} M_{i,t}(M^{m-1})_{t,j}$. But this is precisely $(M^m)_{i,j}$ by the definition of matrix multiplication. This establishes $P(m)$, completing the induction proof.

Thus for each exponent s from 1 to k, the Row i, Column j entry of M^s counts the s-step walks from Node i to Node j. By the definition of matrix addition, the Row i, Column j entry of $M^1 + M^2 + \ldots + M^k$ counts the walks from node i to node j of any length from 1 to k.

15. In each graph, an edge without arrows indicates an edge in both directions.

(a) $\begin{bmatrix} 1 & 0 & 0 & 0 \\ 0 & 1 & 0 & 0 \\ 0 & 0 & 1 & 0 \\ 0 & 0 & 0 & 1 \end{bmatrix}$

(c) $\begin{bmatrix} 0 & 1 & 1 & 0 \\ 1 & 0 & 1 & 0 \\ 1 & 1 & 0 & 0 \\ 0 & 0 & 0 & 0 \end{bmatrix}$

(e) $\begin{bmatrix} 1 & 1 & 1 & 0 \\ 1 & 1 & 0 & 0 \\ 1 & 0 & 1 & 0 \\ 0 & 0 & 0 & 0 \end{bmatrix}$

16. (a) $\begin{bmatrix} 1 & 0 & 0 & 0 \\ 0 & 1 & 0 & 0 \\ 0 & 0 & 1 & 0 \\ 0 & 0 & 0 & 1 \end{bmatrix} \leq \begin{bmatrix} 1 & 0 & 0 & 0 \\ 0 & 1 & 0 & 0 \\ 0 & 0 & 1 & 0 \\ 0 & 0 & 0 & 1 \end{bmatrix}$ is true.

(c) $I \leq M$ is false since $m_{11} = 0$.

(e) $I \leq M$ is false since $m_{4,4} = 0$.

17. For every subscript t, $I_{t,t} = 1$. Therefore $(I \vee M)_{t,t} = 1$ for all t, and hence $I \leq (I \vee M)$. Therefore $I \vee M$ is reflexive.

18. (a) This relation is the same as R in Exercise 15(a).

(c) This relation is $\{(1,1), (2,2), (3,3), (4,4), (1,2), (2,3), (1,3), (3,1), (3,2), (2,1)\}$. The matrix and graph are shown below.

$$\begin{bmatrix} 1 & 1 & 1 & 0 \\ 1 & 1 & 1 & 0 \\ 1 & 1 & 1 & 0 \\ 0 & 0 & 0 & 1 \end{bmatrix}$$

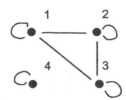

(e) This relation is $\{(1,1), (2,2), (3,3), (4,4), (1,2), (2,1), (1,3), (3,1)\}$. The matrix and graph are shown below.

$$\begin{bmatrix} 1 & 1 & 1 & 0 \\ 1 & 1 & 0 & 0 \\ 1 & 0 & 1 & 0 \\ 0 & 0 & 0 & 1 \end{bmatrix}$$

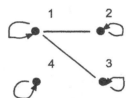

19. (a) $M = M^T$, so this relation is symmetric.

(c) $M = M^T$, so this relation is symmetric.

(e) $M = M^T$, so this relation is symmetric.

21. For parts (a),(c),(e), and (g) the relations (and hence the graphs and matrices) are the same as in Exercise 15.

(b) This relation is $\{(1,1), (1,4), (2,3), (3,2), (4,1), (4,4)\}$. The matrix and graph are shown below.

$$\begin{bmatrix} 1 & 0 & 0 & 1 \\ 0 & 0 & 1 & 0 \\ 0 & 1 & 0 & 0 \\ 1 & 0 & 0 & 1 \end{bmatrix}$$

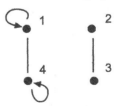

(d) This relation is $\{(1,1), (1,3), (1,4), (2,2), (2,4), (3,1), (3,3), (4,1), (4,2), (4,4)\}$. The matrix and graph are shown below.

$$\begin{bmatrix} 1 & 0 & 1 & 1 \\ 0 & 1 & 0 & 1 \\ 1 & 0 & 1 & 0 \\ 1 & 1 & 0 & 1 \end{bmatrix}$$

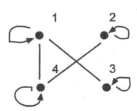

(f) This relation is $\{(1,3),(1,4),(1,5),(2,2),(2,4),(3,1),(3,5),(4,1),(4,2),(4,5),(5,1),(5,3),(5,4)\}$. The matrix and graph are shown below.

$$\begin{bmatrix} 0 & 0 & 1 & 1 & 1 \\ 0 & 1 & 0 & 1 & 0 \\ 1 & 0 & 0 & 0 & 1 \\ 1 & 1 & 0 & 0 & 1 \\ 1 & 0 & 1 & 1 & 0 \end{bmatrix}$$

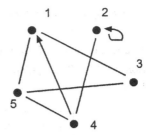

22. (a) $\begin{bmatrix} 1 & 0 & 0 & 0 \\ 0 & 1 & 0 & 0 \\ 0 & 0 & 1 & 0 \\ 0 & 0 & 0 & 1 \end{bmatrix} \le \begin{bmatrix} 1 & 0 & 0 & 0 \\ 0 & 1 & 0 & 0 \\ 0 & 0 & 1 & 0 \\ 0 & 0 & 0 & 1 \end{bmatrix}$ is true.

(c) $\begin{bmatrix} 1 & 1 & 1 & 0 \\ 1 & 1 & 1 & 0 \\ 1 & 1 & 1 & 0 \\ 0 & 0 & 0 & 0 \end{bmatrix} \not\le \begin{bmatrix} 0 & 1 & 1 & 0 \\ 1 & 0 & 1 & 0 \\ 1 & 1 & 0 & 0 \\ 0 & 0 & 0 & 0 \end{bmatrix}$ since $1R2$ and $2R1$ but not $1R1$.

(e) $\begin{bmatrix} 1 & 1 & 1 & 0 \\ 1 & 1 & 1 & 0 \\ 1 & 1 & 1 & 0 \\ 0 & 0 & 0 & 0 \end{bmatrix} \not\le \begin{bmatrix} 1 & 1 & 1 & 0 \\ 1 & 1 & 0 & 0 \\ 1 & 0 & 1 & 0 \\ 0 & 0 & 0 & 0 \end{bmatrix}$ since $2R1$ and $1R3$ but not $2R3$.

23. (a) $\{(1,2),(1,3),(1,4),(1,5),(1,6),(2,3),(2,4),(2,5),(2,6),(3,4),(3,5),(3,6),(4,5),(4,6),(5,6)\}$

(b) The set in the part (a) solution is the transitive closure.

(c) $M^{(1)} = \begin{bmatrix} 0 & 1 & 0 & 0 & 0 & 0 \\ 0 & 0 & 1 & 0 & 0 & 0 \\ 0 & 0 & 0 & 1 & 0 & 0 \\ 0 & 0 & 0 & 0 & 1 & 0 \\ 0 & 0 & 0 & 0 & 0 & 1 \\ 0 & 0 & 0 & 0 & 0 & 0 \end{bmatrix}$

(d) $M^{(2)} = \begin{bmatrix} 0 & 0 & 1 & 0 & 0 & 0 \\ 0 & 0 & 0 & 1 & 0 & 0 \\ 0 & 0 & 0 & 0 & 1 & 0 \\ 0 & 0 & 0 & 0 & 0 & 1 \\ 0 & 0 & 0 & 0 & 0 & 0 \\ 0 & 0 & 0 & 0 & 0 & 0 \end{bmatrix}$

$M^{(3)} = \begin{bmatrix} 0 & 0 & 0 & 1 & 0 & 0 \\ 0 & 0 & 0 & 0 & 1 & 0 \\ 0 & 0 & 0 & 0 & 0 & 1 \\ 0 & 0 & 0 & 0 & 0 & 0 \\ 0 & 0 & 0 & 0 & 0 & 0 \\ 0 & 0 & 0 & 0 & 0 & 0 \end{bmatrix}$

$M^{(4)} = \begin{bmatrix} 0 & 0 & 0 & 0 & 1 & 0 \\ 0 & 0 & 0 & 0 & 0 & 1 \\ 0 & 0 & 0 & 0 & 0 & 0 \\ 0 & 0 & 0 & 0 & 0 & 0 \\ 0 & 0 & 0 & 0 & 0 & 0 \\ 0 & 0 & 0 & 0 & 0 & 0 \end{bmatrix}$

$$M^{(5)} = \begin{bmatrix} 0 & 0 & 0 & 0 & 0 & 1 \\ 0 & 0 & 0 & 0 & 0 & 0 \\ 0 & 0 & 0 & 0 & 0 & 0 \\ 0 & 0 & 0 & 0 & 0 & 0 \\ 0 & 0 & 0 & 0 & 0 & 0 \\ 0 & 0 & 0 & 0 & 0 & 0 \end{bmatrix}$$

$$M^{(1)} + M^{(2)} + M^{(3)} + M^{(4)} + M^{(5)} = \begin{bmatrix} 0 & 1 & 1 & 1 & 1 & 1 \\ 0 & 0 & 1 & 1 & 1 & 1 \\ 0 & 0 & 0 & 1 & 1 & 1 \\ 0 & 0 & 0 & 0 & 1 & 1 \\ 0 & 0 & 0 & 0 & 0 & 1 \\ 0 & 0 & 0 & 0 & 0 & 0 \end{bmatrix}$$ This matches the relation in part (b).

25. (a) The relation is already transitive, so the transitive closure is the same relation.

(c) This relation is $\{(1,1),(1,2),(1,3),(2,1),(2,2),(2,3),(3,1),(3,2),(3,3)\}$. The matrix and graph are shown below.

$$\begin{bmatrix} 1 & 1 & 1 & 0 \\ 1 & 1 & 1 & 0 \\ 1 & 1 & 1 & 0 \\ 0 & 0 & 0 & 0 \end{bmatrix}$$

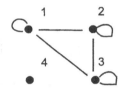

(e) This relation is $\{(1,1),(1,2),(1,3),(2,1),(2,2),(2,3),(3,1),(3,2),(3,3)\}$. The matrix and graph are shown below.

$$\begin{bmatrix} 1 & 1 & 1 & 0 \\ 1 & 1 & 1 & 0 \\ 1 & 1 & 1 & 0 \\ 0 & 0 & 0 & 0 \end{bmatrix}$$

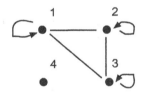

Section 7.5 exercises

1. A solution corresponds to a path from $8,0,0$ to $4,4,0$ in the graph below. Note that in the graph, every node within the square also has edges pointing to two of the four corners of the square, but we have left these edges out for clarity. Note that the puzzle has two solutions, one slightly shorter than the other.

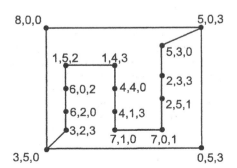

3. The graph model below shows that there is no path from $(10, 0, 0)$ to a node labeled $(5, 5, 0)$. Since the only transitions involve even numbers of quarts and the beginning state has an even number of quarts, we can never measure any odd number of quarts with these containers.

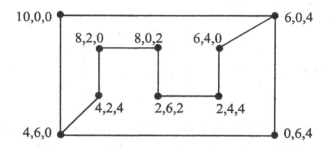

5. Cubes 1, 2, 3 and 4 are shown left-to-right below.

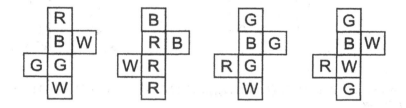

7. (a) The figure below on the left shows the graph of this puzzle, and the graphs on the right show the two good subgraphs. Hence this puzzle can be solved.

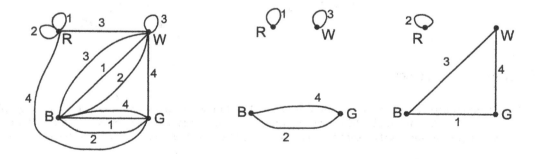

(b) See the figure below. We can argue that the puzzle has no solution as follows. If there *are* two good subgraphs then one will have to include the loop from cube 4 at vertex R while the other must include the edges $[W, R]$ from cube 3 and $[R, G]$ from cube 1. Let's focus on the first of these. The only ways for this subgraph to have every node of degree 2 is to either have B, G, W, B 3-cycle or a B, W, B 2-cycle along with a loop at G. The first of these is impossible since we would be forced to take the $[B, W]$ edge from cube 1 (since cube 4 was already used with the loop at R) leaving us no possible edge $[W, G]$. The second is impossible since there are only two edges of the form $[W, B]$ so we would need to use them both but one is labeled 4 which has already been used with the loop at R. Note that the latter subgraph originally described *is* possible, so there is a partial solution to this puzzle but not a complete solution.

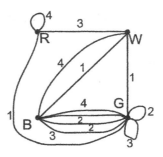

9. In the following grid of states, we follow the procedure outlined:

- Circle $(0, 0)$.
- Cross out (X, X), $(X, 0)$ and $(0, X)$ for all $X > 0$.
- Circle $(2, 1)$ and $(1, 2)$.
- Cross out $(2, X)$ and $(X, 2)$ for all $X > 1$, $(1, Y)$ and $(Y, 1)$ for all $Y > 2$, and $(Z + 1, Z + 2)$ and $(Z + 2, Z + 1)$ for all $Z > 0$.
- Circle $(3, 5)$ and $(5, 3)$.
- Cross out $(5, 7)$, $(4, 6)$, $(3, 6)$ and $(3, 7)$.
- Circle $(4, 7)$.

This leaves $K = \{(4, 7), (3, 5), (5, 3), (1, 2), (2, 1), (0, 0)\}$, Player 1 has the winning strategy. On her first move, she should either remove 1 stone from the smaller pile or 4 stones from the larger pile.

5,7	5,6	5,5	5,4	5,3	5,2	5,1	5,0
4,7	4,6	4,5	4,4	4,3	4,2	4,1	4,0
3,7	3,6	3,5	3,4	3,3	3,2	3,1	3,0
2,7	2,6	2,5	2,4	2,3	2,2	2,1	2,0
1,7	1,6	1,5	1,4	1,3	1,2	1,1	1,0
0,7	0,6	0,5	0,4	0,3	0,2	0,1	0,0

11. (a) This game is identical to the stone removal game played in Exercise 9 but on the grid shown below. Generalizing the process in that solution, we will have a kernel in this game of $K = \{(7, 4), (4, 7), (3, 5), (5, 3), (1, 2), (2, 1), (0, 0)\}$. Since the beginning position has the form $(8, X)$ or $(X, 8)$, it cannnot be in the kernel, so Player 1 will have a winning strategy regardless of the queen's starting spot.

8,8	7,8	6,8	5,8	4,8	3,8	2,8	1,8	0,8
8,7	7,7	6,7	5,7	4,7	3,7	2,7	1,7	0,7
8,6	7,6	6,6	5,6	4,6	3,6	2,6	1,6	0,6
8,5	7,5	6,5	5,5	4,5	3,5	2,5	1,5	0,5
8,4	7,4	6,4	5,4	4,4	3,4	2,4	1,4	0,4
8,3	7,3	6,3	5,3	4,3	3,3	2,3	1,3	0,3
8,2	7,2	6,2	5,2	4,2	3,2	2,2	1,2	0,2
8,1	7,1	6,1	5,1	4,1	3,1	2,1	1,1	0,1
8,0	7,0	6,0	5,0	4,0	3,0	2,0	1,0	0,0

(b) This is like a stone removal game combining the rules of Exercises 9 and 10 and played on the grid shown in the solution to part (a). We can follow the process as before.

- Circle $(0, 0)$.
- Cross out (X, X), $(X, 0)$ and $(0, X)$ for all $X > 0$. Also cross out $(2, 1)$ and $(1, 2)$.
- Circle $(1, 3)$ and $(3, 1)$.

- Cross out $(3, X)$ and $(X, 3)$ for all $X > 1$, $(1, Y)$ and $(Y, 1)$ for all $Y > 3$, and $(Z+1, Z+3)$ and $(Z+3, Z+1)$ for all $Z > 0$. Also cross out $(2, 5)$ and $(5, 2)$.
- Circle $(2, 6)$, $(4, 5)$, $(5, 4)$ and $(6, 2)$.
- Cross out everything else.

Since the beginning position has the form $(8, X)$ or $(X, 8)$, it cannnot be in the kernel, so Player 1 will have a winning strategy regardless of the queen's starting spot.

13. The game is already in a winning position. This means Player 1 will not have a winning strategy assuming that Player 2 has read this section.

$$
\begin{array}{rcccc}
3 & = & 0 & 1 & 1 \\
4 & = & 1 & 0 & 0 \\
7 & = & 1 & 1 & 1 \\
\hline
\text{Direct Sum} & = & 0 & 0 & 0
\end{array}
$$

15. One possibility is piles of 2, 8 and 10 as shown below, but any division of 20 pennies into three piles so that the direct sum is zero will give you the winning strategy.

$$
\begin{array}{rccccc}
10 & = & 1 & 0 & 1 & 0 \\
8 & = & 1 & 0 & 0 & 0 \\
2 & = & 0 & 0 & 1 & 0 \\
\hline
\text{Direct Sum} & = & 0 & 0 & 0 & 0
\end{array}
$$

17. Player 1 should remove all 11 stones in the 8th pile. This will make the direct sum 0.

$$
\begin{array}{rccccc}
1 & = & 0 & 0 & 0 & 1 \\
2 & = & 0 & 0 & 1 & 0 \\
3 & = & 0 & 0 & 1 & 1 \\
4 & = & 0 & 1 & 0 & 0 \\
5 & = & 0 & 1 & 0 & 1 \\
6 & = & 0 & 1 & 1 & 0 \\
7 & = & 0 & 1 & 1 & 1 \\
11 & = & 1 & 0 & 1 & 1 \\
\hline
\text{Direct Sum} & = & 1 & 0 & 1 & 1
\end{array}
$$

19. The complete set of states can be listed in three grids, one for each possible number of stones in the smallest pile.

$$
\begin{array}{llllllll}
257 & 256 & 255 & 254 & 253 & 252 & 251 & 250 \\
247 & 246 & 245 & 244 & 243 & 242 & 241 & 240 \\
237 & 236 & 235 & 234 & 233 & 232 & 231 & 230 \\
227 & 226 & 225 & 224 & 223 & 222 & 221 & 220 \\
217 & 216 & 215 & 214 & 213 & 212 & 211 & 210 \\
207 & 206 & 205 & 204 & 203 & 202 & 201 & 200 \\
\end{array}
$$

$$
\begin{array}{llllllll}
157 & 156 & 155 & 154 & 153 & 152 & 151 & 150 \\
147 & 146 & 145 & 144 & 143 & 142 & 141 & 140 \\
137 & 136 & 135 & 134 & 133 & 132 & 131 & 130 \\
127 & 126 & 125 & 124 & 123 & 122 & 121 & 120 \\
117 & 116 & 115 & 114 & 113 & 112 & 111 & 110 \\
107 & 106 & 105 & 104 & 103 & 102 & 101 & 100 \\
\end{array}
$$

$$
\begin{array}{llllllll}
057 & 056 & 055 & 054 & 053 & 052 & 051 & 050 \\
047 & 046 & 045 & 044 & 043 & 042 & 041 & 040 \\
037 & 036 & 035 & 034 & 033 & 032 & 031 & 030 \\
027 & 026 & 025 & 024 & 023 & 022 & 021 & 020 \\
017 & 016 & 015 & 014 & 013 & 012 & 011 & 010 \\
007 & 006 & 005 & 004 & 003 & 002 & 001 & 000 \\
\end{array}
$$

To find the kernel, we go through the same process as in the earlier problems. We show only an abbreviated process and the final answer here.

- Find the kernel of the game played only on the last grid (the one with no stones in the first pile) just as we did before. The other grids can be ignored during this process. The kernel for this smaller game will be

$$\{000, 003, 006, 011, 014, 017, 022, 025, 030, 033, 036, 041, 044, 047, 052, 055\}$$

- For each kernel state of the form $0XY$ cross out all states of the form $1XY$ and $2XY$ from the other two grids.

- Circle 110 and 101.

- Use the process on the second grid (the one with 1 stone in the first pile) just as we did before. The first grid can be ignored. The kernel states from this grid are

$$\{101, 104, 107, 110, 113, 116, 131, 134, 137, 140, 143, 146\}$$

- Circle 220 and 202.

- Use the process on the first grid (the one with 2 stones in the first pile) just as we did before. The kernel states from this grid are

$$\{202, 205, 220, 223, 226, 232, 235, 250, 253, 256\}$$

This gives a final answer of

$$K = \{\, 000, 003, 006, 011, 014, 017, 022, 025, 030, 033, 036, 041, 044, 047, 052, 055, 101, 104,$$
$$107, 110, 113, 116, 131, 134, 137, 140, 143, 146, 202, 205, 220, 223, 226, 232, 235, 250,$$
$$253, 256\,\}$$

21. The strategy for Player 2 to win is to mirror the moves of Player 1 using the 180° rotational symmetry of the clock face. That is, for each number Player 1 crosses out, Player 2 crosses out the number that is 6 hours later. On each move, Player 1 destroys the 180° symmetry of the picture and Player 2 restores it. Since the winning position has this symmetry, Player 2 must win.

Section 7.6 exercises

2. Answers will obviously vary, but they should be binary trees as long as only biological parents are listed. Here is one possibility:

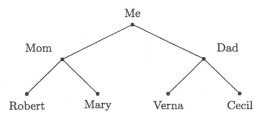

4. Since a complete binary tree with height 3 has $2^3 - 1 = 7$ nodes, we will need to use a tree with height 4 to store 8 values. The following diagram shows one of several possibilities:

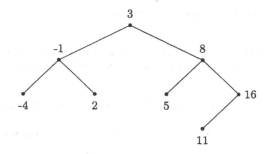

5. Preorder: $ABDEGCF$; Inorder: $DBGEACF$; Postorder: $DGEBAFC$

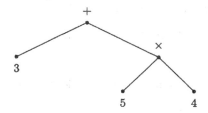

6. (a)

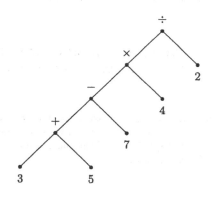

(c)

7. (a) Prefix: $+, 3, \times, 5, 4$; Postfix: $3, 5, 4, \times, +$

    (c) Prefix: $\div, \times, -, +, 3, 5, 7, 4, 2$; Postfix: $3, 5, +, 7, -, 4, \times, 2, \div$

8. (a) $((5 \times 4) + 7) \div 3 = 9$

9. (a) $(6 - (2 \times 3)) \times (12 \div 4) = 0$

10. Inorder: Robert, Mom, Mary, Me, Verna, Dad, Cecil; Preorder: Me, Mom, Robert, Mary, Dad, Verna, Cecil; Postorder: Robert, Mary, Mom, Verna, Cecil, Dad, Me

13. *Proof by induction.* Let $P(n)$ be the statement, "The maximum number of nodes in a binary tree of height n is $2^n - 1$." Statement $P(1)$ refers to a binary tree of height 1, of which there is only one: a single root with empty left and right subtrees. Since such a tree has $2^1 - 1 = 1$ node, we conclude that the first statement $P(1)$ is true.

Now let $m \geq 2$ be given such that statements $P(1), P(2), \ldots, P(m-1)$ have all been checked to be true. Let T be a binary tree with height m. By definition of "height," each of the left subtree T_L and right subtree T_R of T have height less than or equal to $m - 1$. Letting h_L and h_R denote the respective heights

114

of T_L and T_R, we can cite statements $P(h_L)$ and $P(h_R)$, which tell us that tree T_L has a maximum of $2^{h_L} - 1$ nodes and tree T_R has a maximum of $2^{h_R} - 1$ nodes. Hence, the original tree T has a maximum of

$$1 + \left(2^{h_L} - 1\right) + \left(2^{h_R} - 1\right) \le 1 + \left(2^{m-1} - 1\right) + \left(2^{m-1} - 1\right) = 2^m - 1$$

nodes. This confirms statement $P(m)$, completing the induction.

15. Let T be any binary tree. Let L denote the maximum level of any node in T, and let H denote the height of T. Proposition 1 tells us that $L \le H$, and Exercise 14 tells us that $H \le L$. Therefore, $L = H$.

17. (a) The three trees are given below:

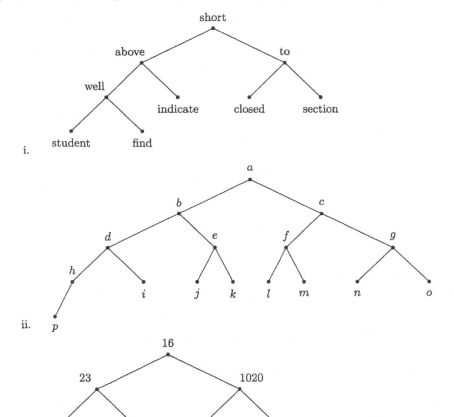

i.

ii.

iii.

(c) 3; 7; 63; 1023

(e) $f(n) = 1 + \lfloor \log_2(n) \rfloor$.

19. *Proof by induction.* Let $P(n)$ be the statement, "$(n!)^2 > n^n$." Since $(3!)^2 = 6$ and $3^3 = 27$, it is clear that statement $P(3)$ is true. Now let $m \ge 4$ be given such that statements $P(3), P(4), \ldots, P(m-1)$ have all been checked to be true. In particular, it has been checked that $((m-1)!)^2 > (m-1)^{m-1}$ is true. From this it follows that

$$
\begin{aligned}
(m!)^2 &= m^2 \cdot ((m-1)!)^2 \\
&> m^2 \cdot (m-1)^{m-1} \text{ by statement } P(m-1) \\
&= m^m \text{ by the given fact}
\end{aligned}
$$

115

Section 7.7 exercises

1. $1, 2, 8, 7, 17, 16, 20, 13, 12, 11, 19, 18, 9, 10, 3, 4, 5, 14, 15, 6, 1$

3. There are $\frac{47!}{2} \approx 1.3 \times 10^{60}$ different cycles to check. At a rate of 10^{15} per second, it would take about 10^{45} seconds, which is about 10^{37} years, far more than the estimated age of the universe.

5. We must have $m, n \geq 2$ since a node of degree 1 puts a stop to a Hamiltonian cycle right away. Note that a Hamiltonian cycle in a bipartite graph has the property that every "step" in the cycle must go from one part to the other. Hence, in order to use every vertex, we must have the same number of nodes in each part. That is, m and n must be equal. It is easy to develop a strategy to traverse any $K_{m,m}$ as long as $m \geq 2$.

6. (a) By considering all 6 possible paths, we find that the cycle a, d, b, c, a has smallest weight 13.

    (b) By considering all 12 possible paths, we find that the cycle a, d, c, e, b, a has weight smallest 24.

    (c) By considering all 12 possible paths, we find that the cycle a, d, b, c, e, a has weight smallest 27.

7. In the graph shown below, if we start with vertex e, the Hamiltonian cycle generated by the algorithm is e, c, d, a, b, f, e and has weight $1 + 2 + 10 + 2 + 3 + 3 = 21$. Note that the cycle a, b, c, d, e, f, a has weight 14, so the algorithm does not give the exact solution to the TSP in this case.

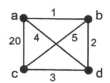

8. (a) The vertex-greedy algorithm produces the cycle a, d, b, c, a with weight 13.

    (b) The vertex-greedy algorithm produces the cycle a, e, b, d, c, a with weight 29.

    (c) The vertex-greedy algorithm produces the cycle a, b, d, c, e, a with weight 28.

9. (a) The edge-greedy algorithm adds edges in the order $[a, d], [a, c], [b, d], [b, c]$, yielding the cycle a, d, b, c, a with weight 13.

    (b) The edge-greedy algorithm adds edges in the order $[a, e], [a, d], [b, e], [c, d], [b, c]$, yielding the cycle a, d, c, b, e, a with weight 24.

    (c) The edge-greedy algorithm adds edges in the order $[a, b], [a, d], [c, e], [b, c], [d, e]$, yielding the cycle a, b, c, e, d, a with weight 28.

13. $[a, e], [a, d], [e, b], [c, d], [c, f], [b, f]$ yields a, e, b, f, c, d, a with weight 42.

14. (a) $00, 01, 11, 10, 00$ is a Hamiltonian cycle in G_2

    (b) $000, 100, 101, 111, 110, 010, 011, 001, 000$ is a Hamiltonian cycle in G_3 shown in the figure below.

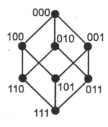

(c) There is a Hamiltonian cycle

$$0000, 1000, 1100, 1101, 1111, 1110, 1010, 1011, 1001, 0001, 0011, 0010, 0110, 0111, 0101, 0100, 0000$$

in G_4

15. (b) The graph on the right below shows the representation of the Knight's Tour on the 4×4 chessboard shown on the left.

 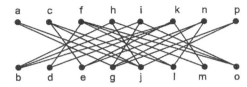

17. The graph shown below has a Hamiltonian cycle, but $\deg(a) + \deg(c) = 4 < 5$.

19. Let G be a simple connected graph on $n \geq 3$ nodes in which each node has degree at least $\frac{n}{2}$. This means that for every pair of non-adjacent vertices u and v, we can be sure that

$$\deg(u) + \deg(v) \geq \frac{n}{2} + \frac{n}{2} = n$$

By Ore's Theorem, G is Hamiltonian.

20. We will refer to three types of edges: five "outside edges," five "star edges," and five "connecting edges." We will assume that the Petersen graph has a Hamiltonian cycle, and argue that a contradiction must result. Since the Hamiltonian cycle starts and stops at the same vertex, it follows that an even number of "connecting edges" must be used. From this and the structure of the graph, we can narrow down the possibilities to three cases:

Case 1. There are two connecting edges, four star edges and four external edges. In this case, all of the star nodes must be visited by a path of length 4 and all of the outside nodes must be visited by a path of length 4, and the endpoints of the first path must be adjacent to the endpoints of the second path. This is impossible since a path of length 4 among the outside nodes must have endpoints that are next to each other in the picture, and a path of length 4 among the star nodes must have endpoints that are *not* next to each other in the picture.

Case 2. There are four connecting edges, four star edges and two external edges. This is impossible because all five external nodes cannot be visited by such a configuration.

Case 3. There are four connecting edges, two star edges and four external edges. This is impossible because all five star nodes cannot be visited by such a configuration.

21. The cycle PA,NY,CT,RI,MA,NH,VT,B,PA has weight 1216 miles.

23. Adding edges

$$[B, VT], [RI, MA], [MA, NH], [RI, CT], [CT, NY], [NH, VT], [PA, B], [PA, NY]$$

in order yields the cycle

$$PA, B, VT, NH, MA, RI, CT, NY, PA$$

with weight 1216 miles.